GIS
For Decision Support and Public Policy Making

Christopher Thomas and Nancy Humenik-Sappington

ESRI Press
Redlands, California

ESRI Press, 380 New York Street, Redlands, California 92373-8100

13 12 11 10 09   2 3 4 5 6 7 8 9 10
Printed in the United States of America

Library of Congress Cataloging-in-Publication Data

Thomas, Christopher, 1963 Aug. 6–
GIS for decision support and public policy making / Christopher Thomas and Nancy Humenik-Sappington.—1st ed.
p. cm.
Includes index.
ISBN 978-1-58948-231-9 (pbk. : alk. paper)
1. Geographic information systems—United States. 2. Decision support systems—United States.
3. Administrative agencies—Information technology—United States. 4. Administrative agencies—
United States—Decision making.
I. Sappington, Nancy, 1948– II. Title.
G70.215.U6T45 2014
352.3'30285—dc22   2008048682

Ask for ESRI Press titles at your local bookstore or order by calling 1-800-447-9778. You can also shop online at www.esri.com/esripress. Outside the United States, contact your local ESRI distributor.

ESRI Press titles are distributed to the trade by the following:

In North America:
Ingram Publisher Services
Toll-free telephone: 800-648-3104
Toll-free fax: 800-838-1149
E-mail: customerservice@ingrampublisherservices.com

Cover design and production   Anthony Fisher
Interior design and production   Donna Celso and Jennifer Hasselbeck
Image editing   Jay Loteria
Editing   Carolyn Schatz
Copyediting and proofreading   Julia Nelson
Permissions   Kathleen Morgan
Printing coordination   Cliff Crabbe and Lilia Arias
On the cover: Capitol Dome photo by Mel Curtis/PhotoDisc/Getty Images. All other photos ©2008 Jupiterimages Corporation.

Contents

Acknowledgments

We are most appreciative to the case-study authors who contributed the articles for this book and who went above and beyond the call, enduring various delays and dealing with our numerous last-minute inquiries. To them, we are gratefully indebted.

There were also many others, including ESRI staff members, whom we wish to acknowledge. They helped us along the way by providing us with data, maps, images, and leads. Keith Mann contributed his ideas to many of the exercises in the book, and in many cases, he single-handedly generated the graphics to support the exercises. Chris Kinabrew also assisted in getting the exercises completed. Most of the exercises in the book were published previously in *Government Matters*. We thank Emily Vines, the editor of that newsletter, for her assistance in getting the information transferred. From outside ESRI, we had help from Brenda and Raymond Martinez with an exercise in chapter 2; and James and Ingrid Ollerton provided the inspiration for an exercise in chapter 3. Mike Heslin, GIS manager for the City of Moreno Valley, California, supported us in our efforts to complete one of the public-participation exercises.

The folks at ESRI Graphics supported us in so many ways with their expertise. We especially thank Kasey Quayle and Amaree Israngkura for their rapid response to our many insistent requests.

Several ESRI business partners pitched in to help in our efforts. Brad Tatham at Scientific Technologies Corp. and the folks at RouteSmart Technologies Inc. assisted us in some of the exercises' graphics presentations.

Finally, we are grateful to Britney Hinthorne and Jeff Allen of ESRI's Industry Solutions government team, who provided us with valuable information on short notice. Their help was especially critical in the administration of all of the permissions materials.

Foreword

September 2008

Policy makers have viewed geographic information system (GIS) technology as a powerful visualization tool since its first implementations. Today, GIS offers real business value with a rich information model and a sophisticated data-management infrastructure. Geography and GIS can provide enormous benefits to an organization, including providing the right kind of support to make accurate and informed decisions.

Government officials are under increasing pressure to make the right choices while minding the budget and delivering value at the same time. You'll find many examples in this book about how they are doing just that. The book can serve as a guide as you look for ways to measure performance and accountability while improving efficiency, customer service, and resource allocation.

Today, thousands of organizations around the world rely on GIS software for supporting a variety of applications. In recent years, GIS server technology has simplified GIS access, opening the door to many new users. These server-based applications are increasingly changing how people do their work, improving communication, collaboration, and decision making. We believe this GIS-based approach of considering all of the geographic factors in a systematic and quantitative framework is valuable not only for improving our organizations, but also for creating a more sustainable world.

Warm regards,
Jack Dangermond
President, ESRI

Return on Investment Matrix

Chapter topics	Cost and time savings	Increased efficiency, accuracy, and productivity	Revenue generation	Enhanced communication and collaboration	Automated workflows	More efficient allocation of resources	Improved access to information
Decision support for budget and finance							
City of Newark, New Jersey	●		●				
City and County of Denver, Colorado	●		●				
Defending a decision/reaching a compromise							
Frederick County, Maryland		●					
City of Houston, Texas		●		●		●	●
U.S. National Park Service		●					
Oneida Nation, Oneida, Wisconsin		●					●
Facilitating public participation in decision making							
City of Cleveland, Ohio		●		●			
Georgia Department of Natural Resources, Brunswick, Georgia	●	●					
Philadelphia GIS Services Group, Philadelphia, Pennsylvania	●	●				●	●
Washington County Planning and Parks Department, West Bend, Wisconsin		●		●			
Making decisions under pressure							
Fort Bend County, Texas	●	●					
Miami-Dade Office of Emergency Management, Miami, Florida	●	●					
Peterson Air Force Base, Colorado Springs, Colorado		●				●	●
Decision support for allocating resources							
City and County of Denver, Colorado	●						
City of Pasadena, California	●	●				●	
U.S. Geological Survey, Menlo Park, California						●	
Utah Department of Health, Salt Lake City, Utah		●				●	●
Making decisions on the fly							
City of Jacksonville, Florida		●			●	●	●
Pottawattamie County, Iowa	●	●					
Zanesville/Muskingum County Health Department, Zanesville, Ohio	●						●
Supporting policies with GIS							
Curry County GIS, Curry County, Oregon					●		
Georgia Department of Human Resources, Division of Public Health, Atlanta, Georgia	●	●		●		●	●
Idaho Department of Environmental Quality, Boise, Idaho	●	●		●	●	●	●
Metropolitan Council, St. Paul, Minnesota	●	●				●	●
Preservation League of New York State, Albany, New York		●					●
City of Sheridan Planning Division, Sheridan, Wyoming		●					
Oak Ridge National Laboratory and University of Tennessee	●	●		●			

Column group heading: Return on investment (ROI) topics

Introduction

As the use of geographic information system (GIS) technology has become more widespread in public agencies and organizations, it is proving to be indispensable in day-to-day business and operations, and its benefits are making a compelling case for using GIS as a strategic tool in decision making. The software is changing how we work and is becoming an essential part of organizational infrastructure where information is embedded in the workflow. It is enabling management to make informed choices and to better assess the impact of their decisions.

How an organization is perceived by the public is based in large part on the assurance its leaders project when rendering a decision. When managers make a decision that is driven by the most up-to-date, accurate data, they can be confident that they have made the best possible choice under the circumstances. The quality of information is expressed in decisions that are timely, relevant, and accurate as well as consistent and objective.

When the buck stops, GIS software can help you quickly compile and integrate information from raw data, documents, personal knowledge, and business models. It visualizes the results, so you can identify and analyze problems, and make the right decisions.

The chapters in this book illustrate how GIS technology is helping people in the public sector carry out policy decisions with confidence and efficiency. In many organizations, the use of GIS software has now become an integral part of decision making with a direct influence on how such decisions are made—and how they are perceived. Whether it involves allocating resources, budgeting, making choices under pressure, supporting policies, or defending decisions, GIS technology is having a profound effect on how agencies function.

An investment in GIS technology has tangible payoffs, whether it is time saved, revenue generated, increased productivity or efficiency, or improved communication. All of these benefits can not only improve performance and accountability, but also build confidence in an organization's ability to make sound, reasoned decisions.

The chapters in this book address the types of decisions officials commonly deal with. Highlighted at the beginning of each article are the kinds of return on investment that have been realized through these strategic decisions. The matrix at the beginning of the book guides the reader to case studies that involve particular types of return on investment.

Decision support for budget and finance

Governments have a responsibility to ensure that taxpayers' money is spent wisely, and the process by which a government budgets and accounts for expenditures such as infrastructure and public services can be complex. With this complexity comes the difficult process of communicating to the public how taxes or rates are determined, setting up the accounting regulations that govern how monies can be spent, and reporting the progress of programs in relation to expenditures. In short, governments are faced with establishing sound, defensible revenue-generating and expenditure policies.

Applying technology to budgeting and finance can be as simple as generating electronic spreadsheets, or as involved as developing complicated accounting packages with the capacity to track and manage all of a department's financial processes. Either way, with the implementation of technology, organizations can realize a significant return on investment.

When governments began factoring geography into the equation, their benefits increased. Those agencies that took a geographic approach to solving and analyzing financial issues inspired the integration of traditional finance packages with GIS. In some cases, the straightforward application of GIS provided new insight into old problems and helped create sustainable financial planning.

Today, it is not uncommon to find GIS at work in cost accounting for developing infrastructure, calculating revenue streams based on geographic elements such as acreage or populations served within a service area, and providing an acceptable accounting methodology to meet federal standards in asset management or fire-service levels. GIS feature extraction and comparison tools provide the means to perform financial audits and reclaim escaped revenues, lending credibility to the idea that GIS is a strong analytical tool.

County assessors use GIS to defend property assessments to taxpayers. Public works professionals use GIS online to communicate where monies are being allocated for capital improvements and to track project schedules and material costs. Economic professionals use GIS to show economic stimulus locations and provide tools to look at investment potential. These uses support the idea that GIS is a strong public participation and communication tool.

Enterprise GIS benefits Newark's many departments

CHAPTER: Decision support for budget and finance
ORGANIZATION: City of Newark
LOCATION: Newark, New Jersey
CONTACT: Christopher Darby, Information Technology manager, Office of Management and Budget gismail@ci.newark.nj.us
PROJECT: Enterprise GIS
SOFTWARE: ArcSDE, ArcIMS
ROI: Cost and time savings; revenue generation

By Christopher Darby

In August 2004, the Office of Management and Budget (OMB) of Newark, New Jersey, began the implementation of its enterprise GIS network. The goal was to create a unified information architecture and solutions environment that would provide the employees of the city and state's public-safety community with easy and quick access to geospatial data and application services. Subsequently, data about each parcel within the city and critical infrastructure, such as schools, hospitals, and transportation hubs, is now readily available to emergency responders. This provides emergency crews with accurate and up-to-the-minute information about an area even before they arrive on the scene.

The Newark Geographic Information Network (NEWGIN) was designed for use by all city agencies. Nearly all of the city's data is linked to the GIS network and can be displayed on a map for analysis. Whether it is from an animal-control officer or a tree inspector, all of the data that city staff collects can be displayed on a geocoded city map and used in the day-to-day decision-making process of city government. This enables the city to use data in nearly all areas of service such as 911 dispatch and response, biohazards, immunization, West Nile virus tracking, tax assessment, building inspections, water-table maintenance, parade routing (traffic), and garbage pickup. The new system provides cost savings through better analysis and strategic planning of how resources are dispatched.

Figure 1.1 GIS data for Newark, New Jersey, is available through NEWGIN.

NEWGIN is built on a platform that includes ESRI's ArcSDE and ArcIMS applications as well as Orion's OnPoint software. The city looks forward to adding Pictometry software to an already developed enterprise system and to enable other municipalities within Essex County to have access to these images. NEWGIN allows information to flow seamlessly from one department to the next.

Keeping traffic moving

The City of Newark has launched a Traffic Advisory Network to be used in conjunction with NEWGIN to provide a comprehensive way to avoid traffic congestion in the city. According to the U.S. Census Bureau, approximately thirty-two million people travel through Newark annually, causing major commuting concerns and high traffic density areas. To help counter traveling through these highly congested areas and to improve travel for work or play, Newark's Traffic Advisory Network monitors high-traffic areas and provides real-time access to traffic patterns on the city's main corridors. Staff can observe traffic flow, congestion, weather, and road conditions to make everyone's journey as safe and efficient as possible. These traffic cameras can be viewed directly within the city's Web-based GIS application, giving users a one-stop shop for all map-related content.

Newark continues to implement new ways of using GIS. In the future, GIS will be useful in driving current and future developmental projects throughout Newark by enabling city staff to use parcel data to identify brownfields, abandoned lots, and lot ownership. This information, when overlaid with other data such as locations of food and beverage establishments, parks and recreation areas, transportation systems, businesses, and other attractions, will provide information to help promote economic growth in Newark. Planned enhancements to NEWGIN include attaching scanned building plans, site plans, floor plans, and any other paper documents to the relevant property. When a user researches a property, all historical information for that property will be viewable.

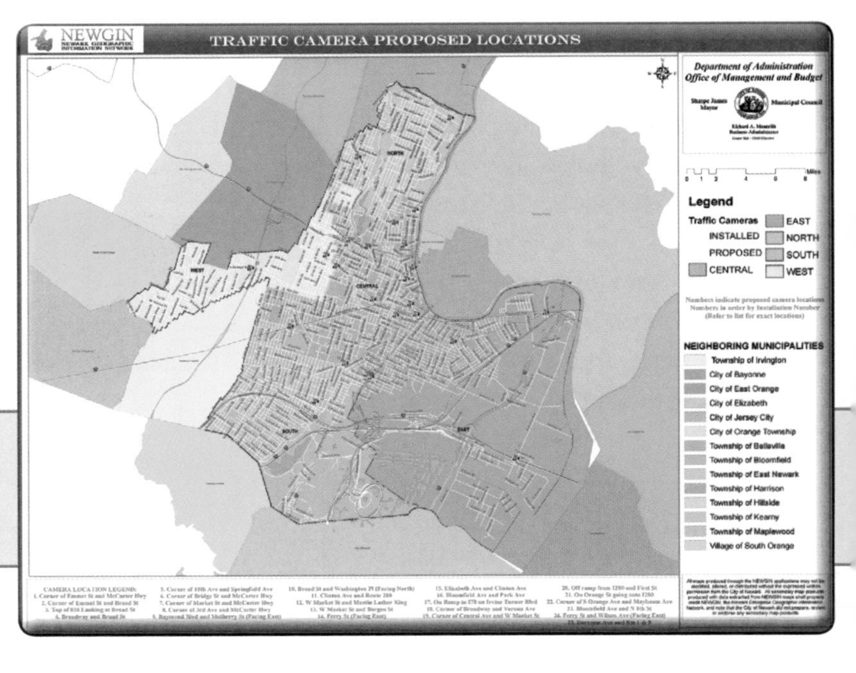

Figure 1.2 NEWGIN map shows proposed locations for the Newark Traffic Advisory Network.

Many benefits realized

Nearly every city department has benefited from the implementation of NEWGIN. The Economic and Housing Development Department was able to pinpoint more than nine hundred properties that have been classified as vacant land and nearly twelve hundred tax-lien city-owned properties. These properties can now be sold to business owners and entrepreneurs to help drive up the city's tax base.

The City Assessor's Office has been one of the biggest users of NEWGIN for a variety of reasons. The Assessor's Office was looking for a better way to determine and print variance lists. Properties within 200 feet of a location subject to renovations were looked up and hand drawn on a map. This process took hours and was especially time consuming if there were a lot of properties on the list. With NEWGIN, a list, a map, a report, and mailing labels can be generated in only three minutes. NEWGIN has freed up more than twenty man-hours per week to be used on other office duties.

The implementation of NEWGIN has helped the city save time in gathering data. It was estimated to take approximately twenty minutes per person, per request to retrieve information from other departments. With NEWGIN, that has been reduced to approximately two minutes. Employees who need assessor tax information can simply type in an address and receive information nearly instantaneously.

The implementation of NEWGIN has helped the city save time in gathering data.

Previously, inspectors sometimes had difficulty locating the owners of a property. With the help of GIS and NEWGIN, the inspectors can call the office, and the information can be quickly relayed. In the future, the inspectors will have tablet PCs with access to the NEWGIN network to look up information themselves, freeing up office crew for other tasks.

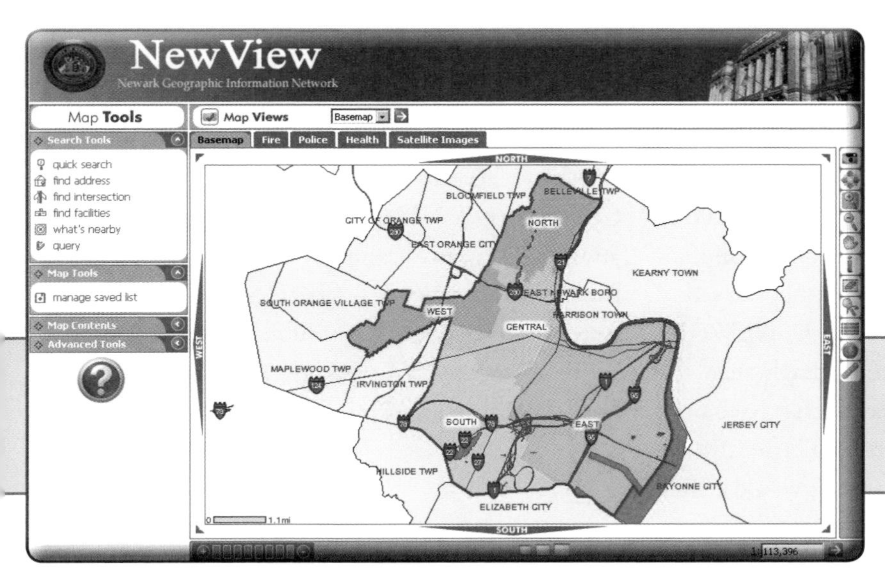

Figure 1.3 The NewView application main page allows easy access to information about Newark.

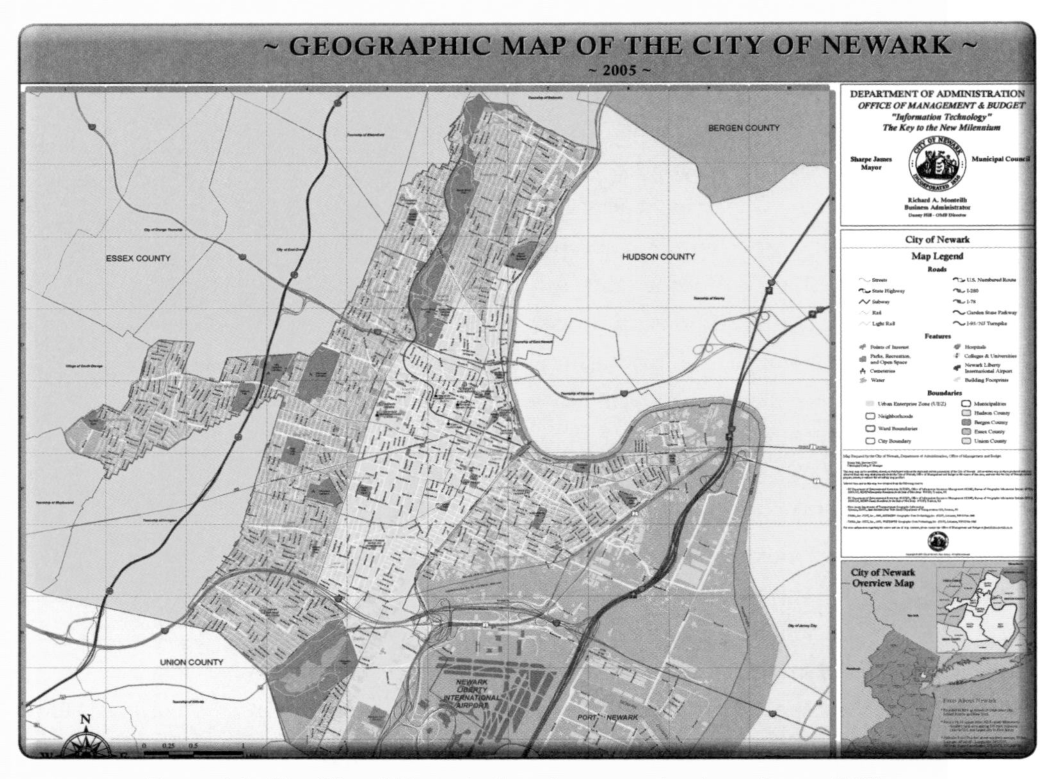

Figure 1.4 The City of Newark, New Jersey, makes broad use of GIS.

Another benefit of NEWGIN is revenue generation. All district maps, such as zoning, election, fire, and police, created under the NEWGIN platform are sold to the public for a fee. A fee schedule has been developed to charge for the production of maps either on CD-ROM or a poster. Before NEWGIN was developed, the city was unable to charge residents for maps because it simply did not have the capability to create enough maps to sell.

Another revenue generator was the NEWGIN support for the citywide auction. Each year, the city hosts an auction of properties that have been foreclosed on by the government. Currently, the city is developing an application that would enable the public to access the location of a house for sale and view photographs of the structure and the surrounding area. This new method will make it quicker, easier, and more informative for buyers looking for property to purchase.

All of the information in the city's enterprise GIS network, NEWGIN, is now available to anyone who needs it, and city employees have more time for other tasks. It has been estimated that the system will have paid for itself in just three years.

To access the NEWGIN network, visit http://njgin.ci.newark.nj.us.

Strict quality control with GIS ensures accurate data for Denver

CHAPTER: Decision support for budget and finance
ORGANIZATION: City and County of Denver
LOCATION: Denver, Colorado
CONTACT: Jeff Blossom, GIS photogrammetry administrator
jeff.blossom@ci.denver.co.us
PROJECT: Orthophoto and planimetric quality control
SOFTWARE: ArcIMS, ArcObjects, ArcView
ROI: Cost and time savings; revenue generation

By Jeff Blossom

In 2004, the City and County of Denver contracted with Sanborn Mapping Inc. to acquire 1:660 scale, six-inch-pixel color aerial orthophotography, and nine planimetric data layers compiled from imagery to keep track of physical resources. The data included impervious surfaces, structures, edge-of-pavement delineation, bridges, parking marks, road stripes, contours, golf course features, and utility features for a 152-square-mile area. The contractor was responsible for recognizing, interpreting, and properly attributing more than 1.5 million features from the imagery. The City and County of Denver was responsible for ensuring that the data delivered was complete, of high quality, and met contract specifications.

The immense volume of data to quality check and the project time frame of 11 months between data delivery and acceptance presented a significant challenge to project-management staff. A common approach to accomplish the quality control (QC) for large data-acquisition projects is to hire a second, independent vendor specifically for QC. This was not an option for Denver because of the lack of adequate funding. Instead, project-management staff decided to recruit and train city employees to perform the QC.

Recruiting personnel for QC was organized through Denver's GIS Steering Committee (GSC). GSC is a multidepartment group of GIS professionals who plan, organize, support, and execute the education, management, and effective distribution of standardized geographic information for Denver. GSC members recruited within their respective departments; participants included Assessment, Public Works, Environmental Services, the Mayor's Office, Community Planning and Development, Safety, Parks and Recreation, Denver Water, and Denver International Airport. One hundred employees were enlisted to perform QC part time.

Orthophoto QC

The contractor developed an ArcIMS Web site for use by all QC personnel. This saved Denver's project-management staff the tasks of developing and deploying an application for each user and allowed QC personnel to do their work anywhere with an Internet connection. The Web site included standard pan/zoom functionality and the ability to enter an error point with a comment. Training sessions were conducted to educate all QC personnel on how to use the Web site, and when to cite a photo error.

Project-management staff developed training manuals and curriculum with the intention of achieving a consistent level of QC, regardless of the person checking, or the area of Denver being reviewed. Photo errors were grouped into four types: (1) structure lean and warping, (2) scratches and artifacts, (3) mosaic errors, and (4) brightness, shadows, and discoloration. Several examples of each error were shown during training, using existing imagery datasets. Common misinterpretations such as power lines that could appear as scratches, as well as acceptable shadowing and natural ground discoloration were also demonstrated during training.

Imagery was tiled into 1/16-mile squares (1,320 feet by 1,320 feet), producing 2,433 tiles. For estimating purposes, project-management staff met and reviewed several image tiles of varying detail to derive an average review time per tile and a recommended map scale for reviewing. The group decided on a review scale of 1:400, and an average review time of seven to ten minutes per tile. The fact that most of the QC personnel had limited experience with orthophoto review was also included in deriving the time estimation.

The city was split into twelve delivery areas for the project. Area 1 imagery was made available online for QC in August 2004. Subsequent areas were made available in pairs approximately every two weeks, ending with Area 12 in December 2004.

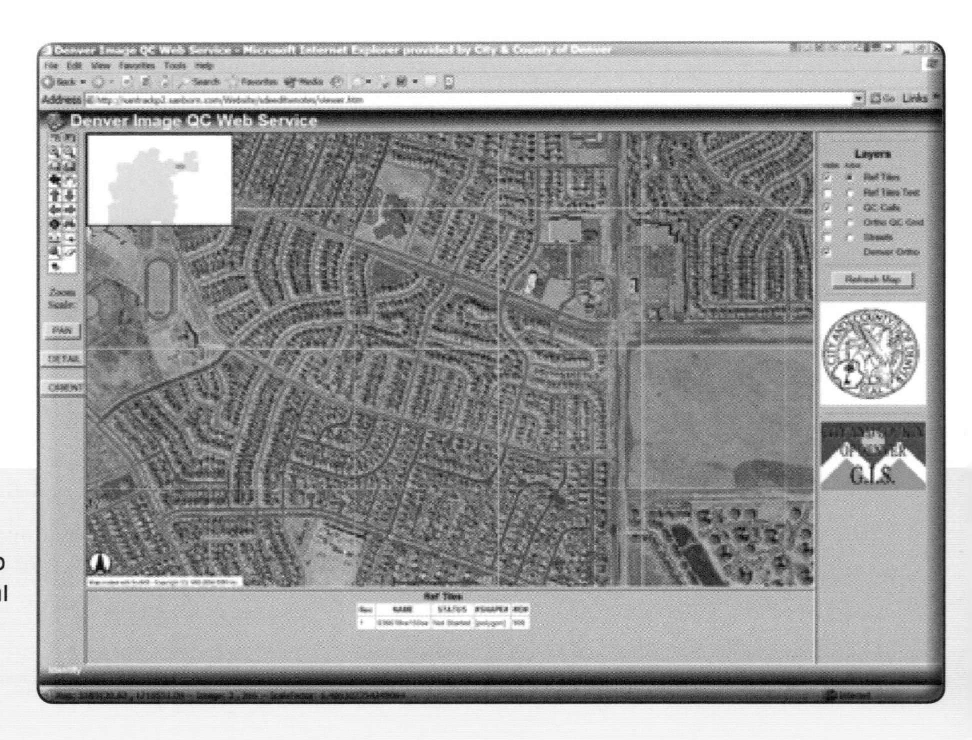

Figure 1.5 ArcIMS orthophoto QC application uses color aerial photography to keep track of Denver's natural resources.

Each department coordinator submitted a letter detailing how much time and personnel they could commit to the QC effort during the five-month period. The goal was to finish QC on an area as the next area was delivered. Using this estimation and availability information, image tile lists were created and distributed to the various departments.

Photo interpretation is not an exact science, and this became strikingly apparent as soon as the project began. Project management was shocked at the volume of invalid photo-error calls made by QC personnel during the first few days of live QC. Errors were called for features such as tilled soil (scratch error), tree shadows, oil stains on streets, cracks in sidewalks, minor building lean, skylights, stairs (mosaic error), and even control panels. It seemed as if some people were under the impression that the more errors they called, the better they were doing. It was clear that project-management staff had underestimated the impact on QC consistency and accuracy by using novice photo interpreters. The two-hour training session apparently was not comprehensive enough to achieve the desired level of consistency among QC personnel. Project management responded with an e-mail to all QC personnel urging them to call errors strictly according to the guidelines in the training manual. Fortunately, the manual was detailed, so that this was possible. Included in the e-mail were screen shots of invalid photo-error calls, additions to the list of commonly misinterpreted features, and specific examples of what not to call.

This immediate response from project management helped decrease the amount of invalid error calls, but a review of each error was still necessary to filter out invalids before sending the errors on to the contractor. A GIS technician was delegated four to eight hours a day to filter out invalids under the project manager's close supervision. Training for the October–December QC personnel was scheduled during September, so the lessons learned during live QC in August were incorporated into these training sessions. Strict adherence to the training manual was also emphasized during subsequent training sessions. After examining the Area 1 imagery, it was clear that the imagery was of high quality and few valid photo errors actually existed; this information was passed on to QC personnel as well. Explaining that most tiles probably would not have errors also helped reduce invalid error calls. The orthophotography QC progressed smoothly after the initial few weeks. Valid errors were returned to the contractor, fixed, and final imagery was delivered on time in January 2005.

This immediate response from project management helped decrease the amount of invalid error calls.

Planimetric data QC

Quality control for the planimetric data began in October 2004 and lasted through June 2005. Project management's strategy was slightly different for this phase of the project, leaving most of the recruitment to the department responsible for the data. For example, Public Works was responsible for checking the edge-of-pavement layer, Parking Management for the parking stripes layer, and so on. Project management handled the QC of nonspecific department layers such as contours and structures. When the various departments defined QC personnel, project management handled the QC training and error submittal to the contractor.

To perform the QC, the project management team used ArcObjects to develop an application for ArcView. Upon opening the application, the user was prompted to enter his/her name, layer to QC, and tile number. The application then zoomed in on the tile and displayed the imagery, appropriate layers, and error call toolbar. Using the orthophotos as a reference, the planimetric data was checked for the following error types: missing feature (feature present on the photo, but not captured); geometry error (feature's shape incorrectly collected); misclassification (incorrect identification entered in attribute table); spatial error (feature collected in wrong geographic position); and phantom error

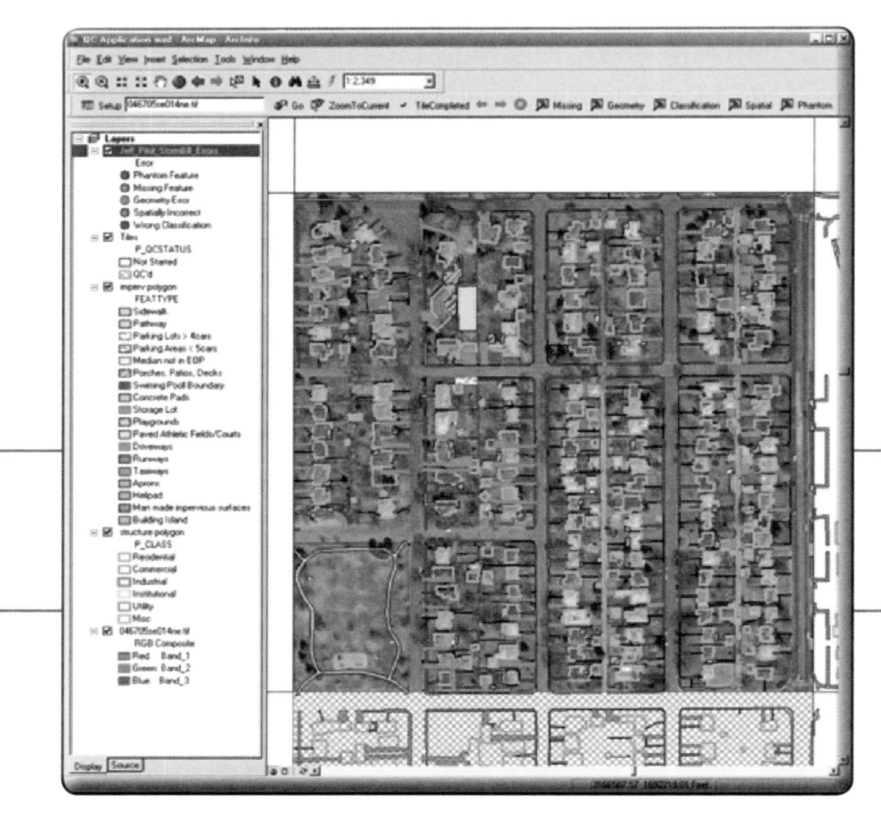

Figure 1.6 Custom ArcView planimetric data QC application is used to check for data-entry errors.

(feature collected does not exist on the photo). The project-management team reviewed all error calls before sending them back to the contractor.

Again, consistent photo interpretation was an issue, but this time between project management and the contractor. Disputes about what constituted a valid error were resolved during a visit by the project-management team to the contractor's work site. The main issue was the contractor's inadequate collection of impervious surfaces in shadowed areas. During stereoscopic compilation, the contractor did not collect sidewalks and driveways that were clearly visible on the orthophotos. The reason was revealed after a laptop was set up to display the orthophotography next to the soft-copy stereoscopic compiler workstation. The same problematic shadowed areas were displayed on both machines for comparison. This revealed that many of the ground features visible on the orthophotos were more difficult to identify on the soft-copy system. The orthophotos are a mosaic of the best imagery available for each area, color balanced and displayed on a monitor that can be fine-tuned for imagery display. In the soft-copy system, two images (not necessarily color balanced, or the best visual images for each area) are displayed through different channels for 3D viewing. This does not allow for the optimum visualization provided by the orthophotos. This viewing difference, combined with the contractor's standard of only collecting features that were positively identifiable (i.e., not guessing), is what caused the discrepancy. The project-management team now understood why features in shadowed areas were being under-collected and incorporated this new information into the QC process. The project then moved on smoothly and finished on time in June 2005.

Plan reaps benefits

Denver included several specifications in the contract that ended up paying significant dividends. Enforcing a 99.99 percent accuracy rate for all data deliverables emphasized the importance of an exceptional mapping standard for the contractor and forced Denver to be extremely detailed with QC from the beginning. Requiring the contractor's project headquarters office to be located within seventy-five miles of the city (Sanborn is in Colorado Springs, sixty miles away) was a contract specification that enabled the two parties to meet frequently at the

Enforcing a 99.99 percent accuracy rate for all data deliverables emphasized the importance of an exceptional mapping standard.

appropriate site to resolve project details. Next time, inclusion of more thorough orthophoto QC training will be employed, perhaps including a pilot area QC zone for participants. The issue of stereoscopic versus monoscopic feature identification in shadowed areas could be resolved by adding a specification that the contractor perform feature collection on the final orthomosaic in addition to the stereoscopic compilation. However, this addition would add significantly to project cost.

Performing QC for a large-scale acquisition of orthophoto and planimetric data is a task that should not be taken lightly. Denver was able to commit large portions of time for project-management staff, which fluctuated between three and six GIS professionals throughout the project. An additional ninety-four staff members composed of engineers, investigators, plan-review technicians, field personnel, clerks, administrators, and appraisers contributed from twenty to two hundred hours each for the QC. Through this effort, combined with Sanborn's high level of quality and professionalism, Denver now has an unprecedented amount of highly accurate orthophotography and planimetric data.

The City and County of Denver was able to save money by not hiring a QC vendor. Acquiring updated impervious surfaces for storm-water billing and then reselling that imagery and data also generated revenue for Denver.

The City and County of Denver was able to save money by not hiring a QC vendor.

Exercise in decision support

Not letting city services get snowed under

Perhaps the most contested decisions that government executives and elected officials make revolve around how a jurisdiction spends taxpayer dollars. Generally, local newspapers cover the annual budget process, citizens attend hearings to express their concerns, and government departments come armed to do battle in securing the funding they need to successfully run their programs. The budget process provides a forum where departments can vie for scarce resources and present their cases to elected officials and budget officers. Each program by itself appears worthy, but the reality is that not all efforts can be funded.

After all programs and their associated costs are presented, the process of prioritizing fund allocation to those programs begins. Decisions are made, and a final budget is adopted. The elected bodies are held accountable for the programs they fund, while department managers prepare staff to deliver services within a limited budget. It is possible that staff members will have to deliver the same levels of service without the aid of additional staff and/or new equipment, or they might not have their programs funded at all.

So, what happens when all projected service levels and budgets fall short because of outside forces? This exercise will look at using GIS technology as an alternative to cutting programs or reducing service levels.

Rising gas prices and bad weather

Governments must manage fleets of field workers, who are involved in services such as collecting trash, repairing potholes, removing graffiti, plowing snow, and sweeping streets. One can easily defend the position that trash must be picked up regularly and snow-covered streets need to be cleared to keep the economy and people moving.

In the following scenario, a public works department faces major budget overruns and must set aside the plans it has made based on experience, projected growth, and a fairly accurate handle on estimated costs. Yet the region has witnessed record snow levels. Staff has run into extreme overtime costs working around the clock to clear the streets. Now, to make matters worse, gas prices have topped a record $4 per gallon.

The public works department continues to provide the services expected of it, which helps to keep the community mobile. Inevitably, elected officials and department heads are faced with the dilemma of dealing with budget overruns. These budget shortfalls are a problem that every department has to address collectively.

Elected officials have asked management to present an adjusted budget. One would expect that departments will need to reduce spending on other programs or face the decision to cut or delay projects indefinitely. If you were in this predicament, what would you do? Is there an alternative to cutting programs while maintaining adequate levels of service?

The GIS difference

An amended budget with significant cuts would probably be the first step government executives and elected officials would consider when reacting to the operations and capital overruns. However, if the leadership asked for alternatives to budget cuts, would you suggest that GIS technology could assist in dealing with this problem?

What role can a GIS play in adjusting service delivery? The answer lies in looking at the various routing solutions that GIS technology provides. At the core of routing and networking solutions is the ability to move fleets or individuals on an optimal route from point A to point B. This ability provides time and fuel cost savings for personnel.

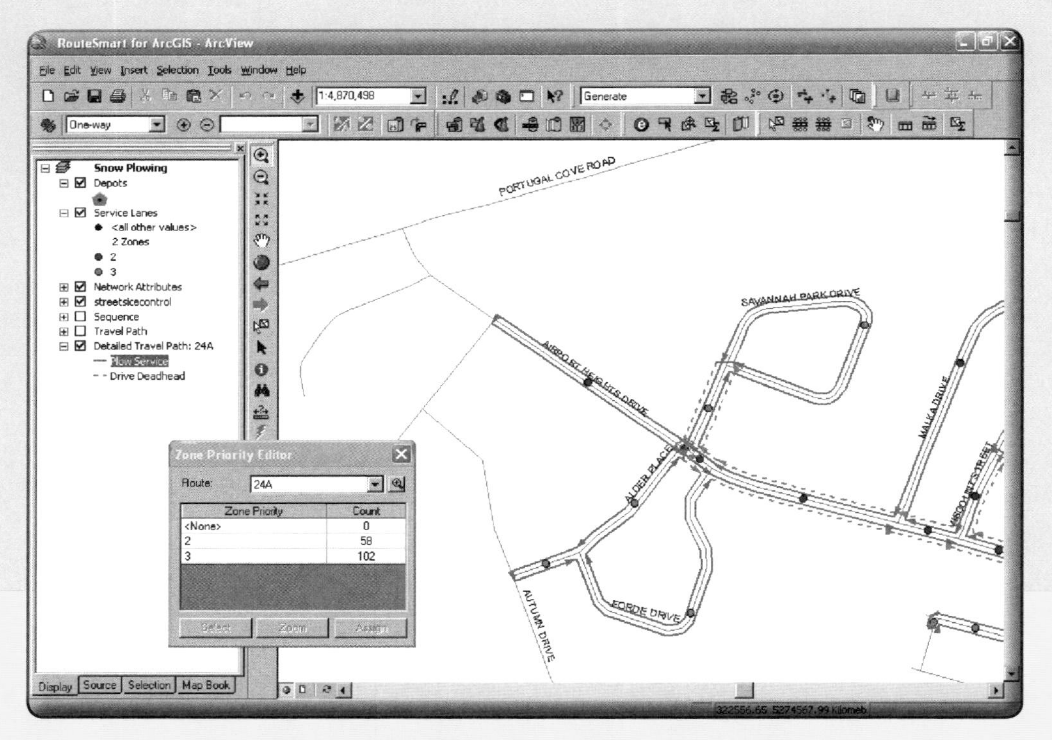

Figure 1.7 The service area is partitioned into routes to be plowed. Priority is maintained for major arterials and emergency routes.

Since this exercise included increasing fuel prices and the need to adjust service levels based on bad weather, the enhanced capabilities of GIS to assist staff in adjusting their scheduling should be noted. The scalable solutions for routing can help manage constraints such as new commercial or residential developments; designated time windows; increased workloads and achieving workload balances; personnel issues such as vacations and illness; and equipment issues. In nearly every case where GIS technology has been deployed, sites have seen significant savings in personnel, vehicle, fuel, and material costs.

The same solution that is applicable to the snowplow problem can be used for street sweeping, solid-waste pickup, recycling pickup, meter reading, street maintenance, and graffiti abatement.

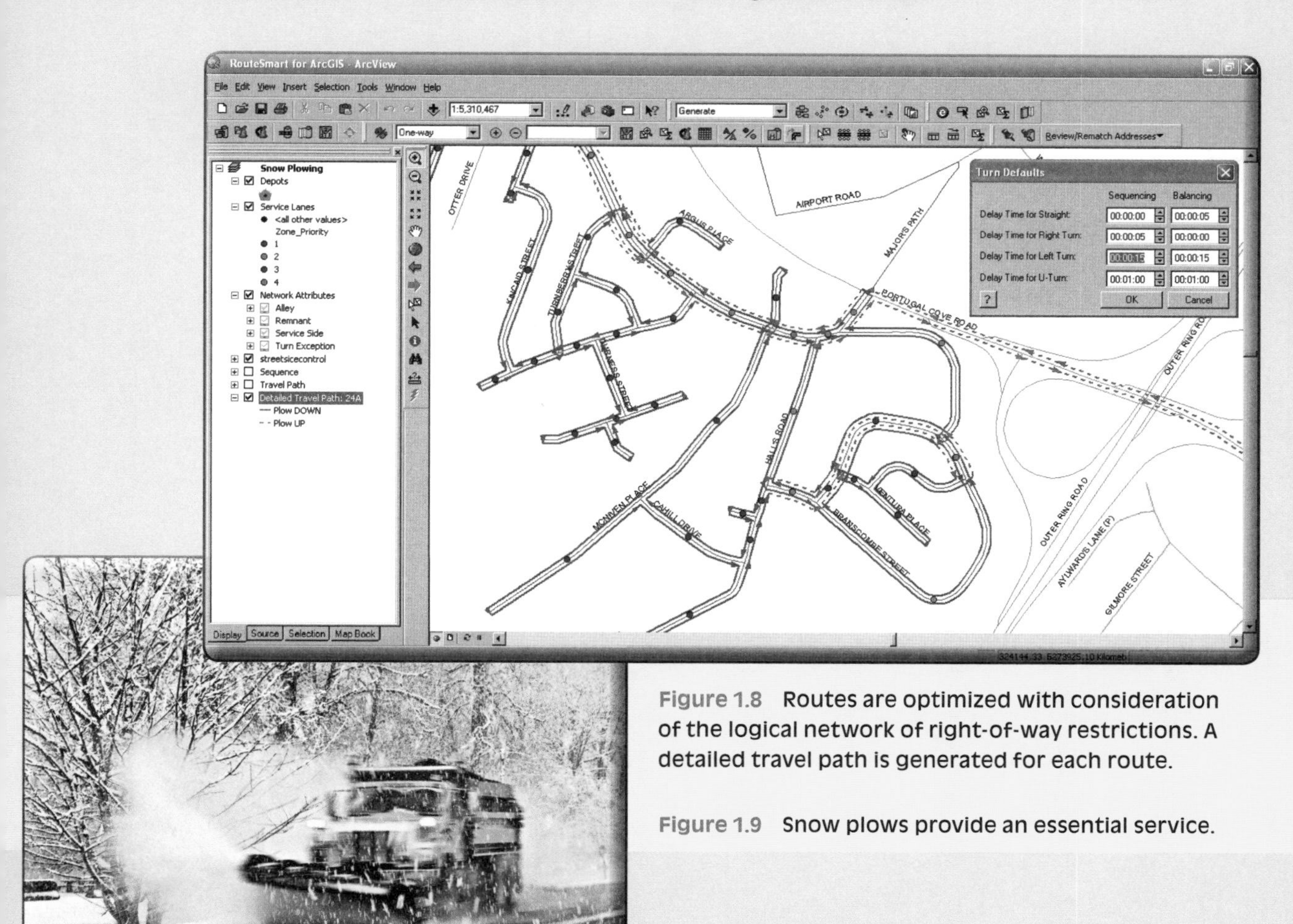

Figure 1.8 Routes are optimized with consideration of the logical network of right-of-way restrictions. A detailed travel path is generated for each route.

Figure 1.9 Snow plows provide an essential service.

Defending a decision/ reaching a compromise

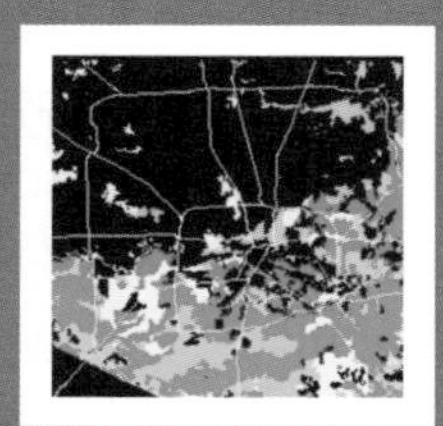

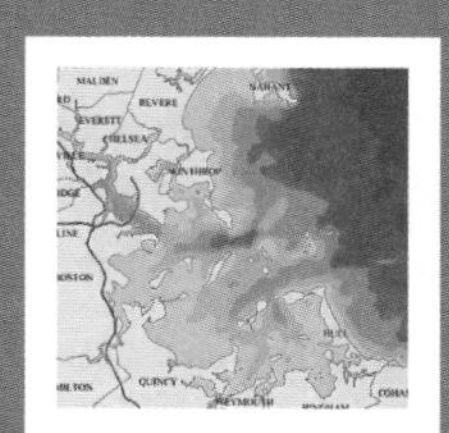

2

Developing sound public policy often involves reaching a compromise between individuals or parties who look at the world from varying perspectives. These differences could arise from a variety of factors—from a disciplinary or professional orientation to protectionist or conservationist concerns to prohibitive views, or simply from personal gain.

Consensus decision making differs from the method where a decision maker looks at various alternatives and makes a choice based on the path of least resistance. Public policy derived from consensus looks for a decision that serves as an alternative. The process seeks to solve problems and avoid conflict.

Decision making through compromise does not necessarily look to develop a policy in which the result is a win-win solution for all. Rather, this practice aims to enable parties with different viewpoints to feel as if they can move forward, having fulfilled their sense of purpose. Through consensus, government officials are not only working to achieve the best solution, but also to promote the growth of the community and foster its trust.

Examples of public policy by consensus might involve developing a balance between a developer and environmental groups, allowing certain types of First Amendment rights to coexist with religious convictions, or not allowing one interest's financial objectives to interfere with a community's self-image.

GIS technology can support this method of developing public policy in several ways. The most obvious is to support visualization of alternatives, but it could be used even better to form the consensus proposals. In this way, the technology serves to diffuse emotions that sometimes drive public policy. Stakeholders can use the GIS to illustrate the points they hope will be considered. Incorporating these points can further decision making by modeling an "if this, then that" scenario. For example, the software could be used in the process of determining where to place wind turbines. A biologist who might not support wind energy but would prefer solar or nuclear alternatives could demonstrate that a proposed site is also the migration path of a specific bird or insect, and the proposed corridor could cause casualties to the species. The GIS could then be used to locate an alternative high-wind corridor outside the migration path.

The role of GIS in consensus decision making provides for qualitative over quantitative analysis. GIS can set aside decisions argued with emotions by presenting facts and alternatives where both parties are technically correct but can come to a mutual agreement. Moreover, the technology can provide visual, analytical, and demonstrative qualities that can accommodate all interested parties and present facts without emotional overtones.

Mulling controversial development in Frederick County

CHAPTER: Defending a decision/reaching a compromise
ORGANIZATION: Frederick County, Maryland
LOCATION: Frederick, Maryland
CONTACT: Amber DeMorett, GIS project manager II ademorett@fredco-md.net
PROJECT: Comprehensive Region Plan Update
SOFTWARE: ArcMap, ArcIMS, ArcReader
ROI: Increased efficiency, accuracy, and productivity

By Amber DeMorett

Frederick County is the gateway to Maryland's western panhandle, bordered by Pennsylvania to the north and Virginia to the south. In the mid-1970s, Frederick County was a growing farming community with little residential or commercial development. The landscape changed during the 1980s when the county experienced a population boom, which continues today. Much of the residential development pressure stems from relatively low housing costs and the county's proximity to Washington, D.C., and Baltimore. The development pressures transforming this once robust farming community into a more urbanized landscape are not likely to dampen in the future.

Comprehensive plans are essential in areas experiencing great residential and commercial development stresses. The policies and recommendations put forth in these planning documents can guide the characteristics of development and meet the desires of current residents. Well-thought-out comprehensive planning accommodates new population growth and development by enhancing the existing community's identity, protecting the environment, creating economic opportunities, providing infrastructure, and aiding in preserving long-range considerations in short-range decision making.

For planning purposes, Frederick County is divided into eight planning regions that periodically undergo a Comprehensive Region Plan Update. One of these planning regions, New Market, is slated for a large portion of the county's growth and has been subject to much controversy. The update process for the New Market Region Plan began in March 2001 and was not adopted until May 2006. The 2006 region plan, at build-out, projected approximately fourteen thousand new housing units. This plan was not supported by the public and became a campaign issue for the Board of County Commissioners (BOCC) elections in fall 2006. The winning candidates were elected partly because of a campaign platform aimed at revising the 2006 New Market Region Plan.

When the new BOCC took office, members initiated a New Market Region Plan Reconsideration of the 2006 plan, which they would prepare themselves. Because of the controversy and undefined processes surrounding this atypical review, BOCC members needed to justify their decisions to the public and quickly become familiar with changes that occurred between region plan iterations, the physical landscape, and regional development potential. They needed parcel-based data on future land use, environmental features, current zoning, and tax assessments as well as the supporting orthophotography. They wanted to learn as much as possible about a property to be able to create a more comprehensive development outlook.

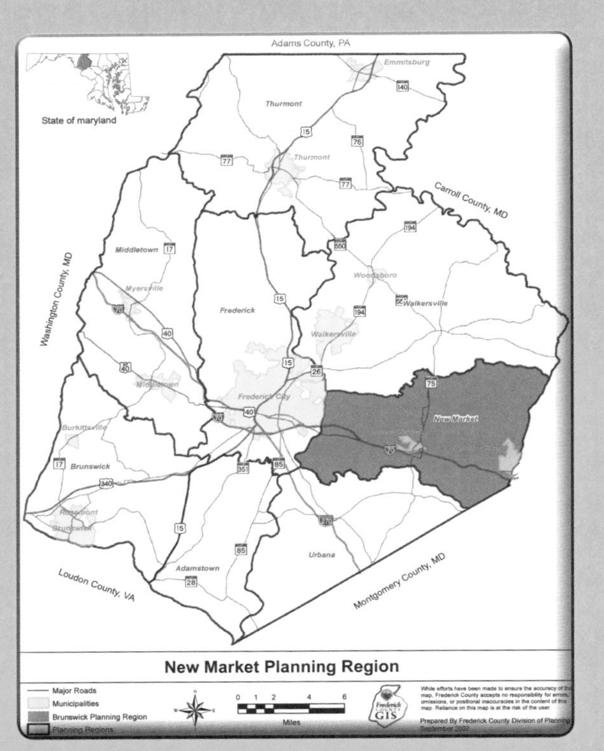

Figure 2.1 Locator map shows the New Market Planning Region in Frederick County, Maryland.

A tool for visualization

GIS technology was an invaluable tool to commissioners for gaining an enhanced understanding of the physical landscape, current land-use designations, and development potential and impacts within the New Market Region. The GIS provided commissioners with instantaneous visualization of changes and their effects on overall development. Before implementing the GIS in the reconsideration process, one main issue needed to be addressed. The BOCC, comprehensive planning staff, and GIS staff all had various levels of exposure to and understanding of how a GIS could be utilized as a powerful planning tool in the region plan update process.

The newly elected BOCC members had a broader understanding of technology than their predecessors; however, they had not been directly exposed to the software. They were familiar with it through end-user products such as hard-copy maps and/or spreadsheets. At the beginning of this planning process, they had many questions about GIS technology and its capabilities.

The comprehensive planning staff was more familiar with the software and had direct exposure to it and its powerful analysis capabilities. They had access to ArcReader, and many of them used it in daily decision making. They were not as familiar with the efficiency and ease of data manipulation, area calculations, visualization, database management, and the more robust analytical features of GIS technology. The planning staff mainly used the software for midlevel end products such as hard-copy maps and ArcReader applications. The GIS staff had extensive GIS experience and knowledge within the field; however, many of them did not have exposure to planning processes and theories. Many of the daily activities for the GIS staff are to generate improvements to the Division of

Planning's workflow. Mostly, planners ask GIS staff if certain tasks or ideas are possible with the software, and the GIS staff transforms these ideas into working planning tools.

Before the New Market Region Plan Reconsideration, GIS technology was not used for planning region updates other than creating hard-copy maps for public hearings and report documents. During the region updates, a new land-use plan map is adopted that shows future land-use designations and is used as a development guide for properties during a twenty- to thirty-year period. Revisions to previous plans are recommended by the planning commission and adopted by the BOCC. Typically, the process of revising the maps involves keeping track of changes made at public workshops, hand drawing the changes on existing basemaps, reprinting the maps, and finally reviewing the maps to make sure changes have been made correctly. This inefficient cycle of changes, which creates more opportunities for data errors and omissions, is repeated after each workshop until a final land-use plan map is adopted.

Revisiting the old plan

The BOCC began the New Market Region Plan Reconsideration at a public workshop in January 2007, where it reviewed and discussed New Market region plans dating back to 1984 and 1993, deliberated about how to proceed with this atypical update, and determined which of the previous region plans should serve as a starting point. Two land-use scenarios were presented. One detailed subdivision developments that were already approved. The other scenario, presented by one of the commissioners and coined the "Geronimo Option," showed an entirely new set of zoning and comprehensive land-use maps.

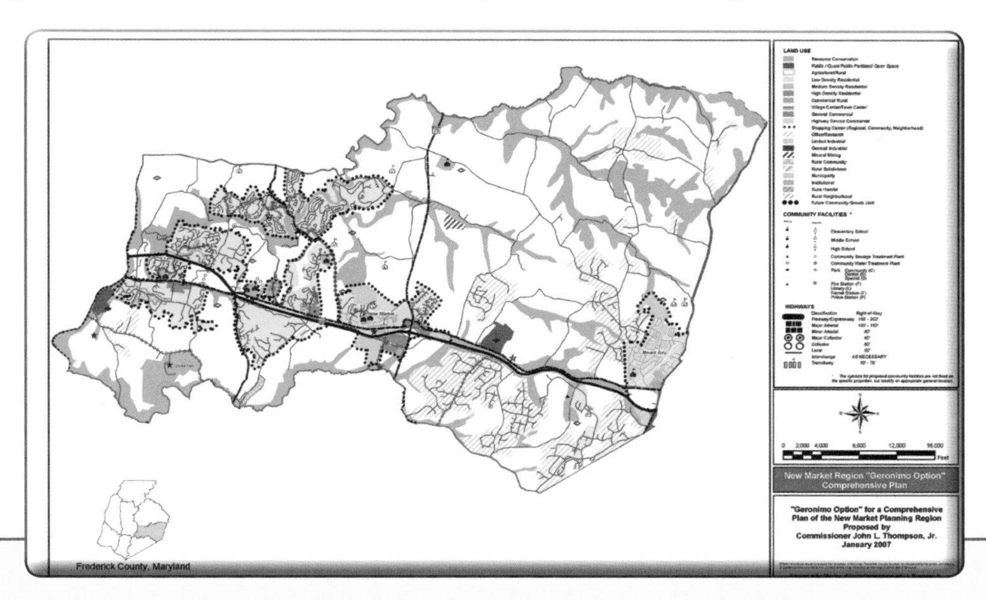

Figure 2.2 The final "Geronimo Option" represents the first collaboration between an elected official and GIS staff.

The commissioner worked with GIS staff to visualize his ideas for other BOCC members and the public. Within days, staff had created a new set of land-use planning and zoning maps that accurately depicted his ideal development scenario, along with an Excel file of land-use designation acreage changes that would occur under the Geronimo Option.

GIS technology was implemented at the BOCC workshops as a way to display all of the different development scenarios and iterations of the New Market Region Plan. After an initial workshop where only ArcReader was used, the GIS staff switched to an ArcMap application so that staff could perform on-the-fly-editing, query data, add and manipulate GIS layers, and create comprehensive visuals.

Public workshops began with display of the basemap. GIS staff made changes as requested during the workshop, and zoomed in on areas for more detailed consideration. The area of concern was defined and digitized, creating an accurate boundary of change. As the BOCC discussed the area of concern, commissioners would ask staff to overlay various layers such as current zoning, floodplains, forest cover, steep slopes, and orthophotography that would potentially affect their decisions. Using this process, the BOCC could demonstrate to the public why it was making certain land-use changes. The use of GIS technology also aided in controlling the discussion and enabled the BOCC to contemplate the issues and potential implications for development of an area. ArcMap became an integral part of the reconsideration process. The BOCC was impressed with the efficiency of on-the-fly editing, instant-change

December 13, 2006

GIS Estimate of Acreage Change in New Market Planning Region from the 2006 Adopted Zoning and the "Geronimo Option"

Zoning District	Current Zoning (1)	"Geronimo Option"	Acreage Change
A	30,699.6	34,393.9	3,694.3
GC	271.4	238.6	-32.8
GI	176.9	5.1	-171.8
HS	2.6	2.6	0.0
LI	302.3	175.9	-126.4
MM	52.3	52.3	0.0
MUN	1,332.6	1,332.6	0.0
ORI	99.3	0.0	-99.3
PUD	4,752.7	2,571.4	-2,181.3
R1	6,733.4	5,892.4	-841.0
R3	727.9	600.0	-127.9
R5	113.5	0.5	-113.0
RC	1,650.3	1,650.3	0.0
VC	2.2	1.4	-0.8
Total	46,917.0	46,917.0	

Rough GIS acreage estimates of change.
GIS Estimate of Acreage Change in New Market Planning Region from the 2006 Adopted Zoning and the "Geronimo Option"
(1) As adopted with the New Market Region plan, May 2006

Table 2.1 Table shows the land-use acreage changes that would result if the "Geronimo Option" were adopted.

visualizations, and the ability to defend its changes based on the environmental features of a property.

Return on investment

This strategy saved time by involving GIS technology directly in the planning process. The BOCC could get instant answers to environmental and developmental questions on a parcel-by-parcel basis and see how land-use designations would affect specific properties. Planning staff members no longer had to rely on hand-drawn maps to depict their ideas. The BOCC, staff, and the public could follow together and see how the proposed changes would affect this controversial area. Through the GIS software, the BOCC also had the ability to let people know and understand why they were making certain land-use decisions.

Through the GIS software, the BOCC also had the ability to let people know and understand why they were making certain land-use decisions.

After a solid strategy was developed on how to use GIS technology, the software became an integral part of the decision-making process. The on-the-fly editing capabilities provided assistance in decision making and instantaneous visualization throughout the process. Without the use of ArcMap, this process could not have gone as smoothly. Since its interaction with GIS technology, the BOCC has come to rely on GIS staff, asking it for assistance with other issues in the county. The enterprise GIS staff has created an ArcIMS site especially for the BOCC, so members can use this Web application to view base data anywhere and anytime the need arises. The BOCC has begun to fully understand how its monetary investment is paying for itself through efficient decision-making processes and visualization.

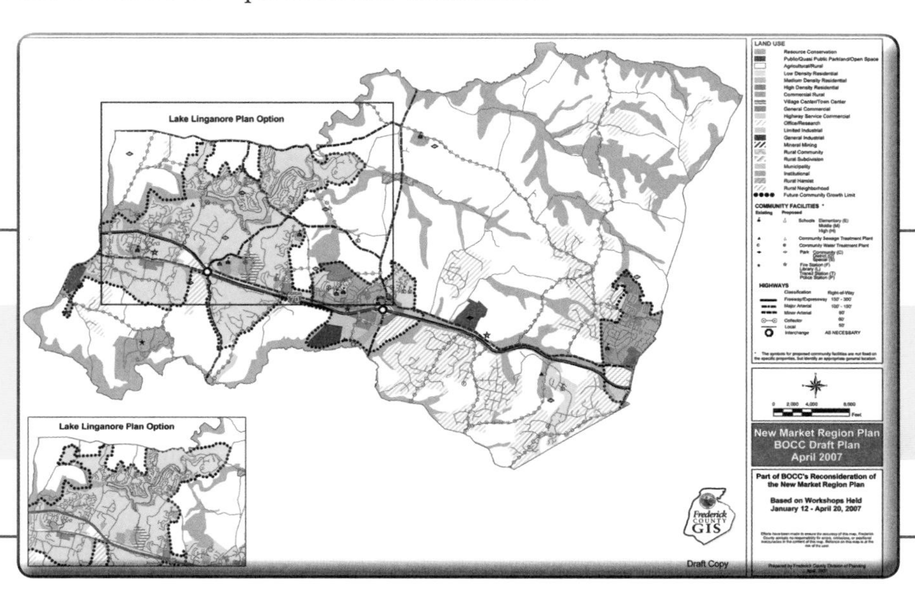

Figure 2.3 The final New Market Region Plan Reconsideration prepared by the BOCC used GIS to justify rezoning.

What to fix first in the twenty-first century

CHAPTER: Defending a decision/reaching a compromise
ORGANIZATION: City of Houston
LOCATION: Houston, Texas
CONTACT: Brian Long, supervising engineer brian.long@cityofhouston.net
PROJECT: Water Infrastructure Replacement Prioritization
SOFTWARE: ArcGIS
ROI: Increased efficiency, accuracy, and productivity; enhanced communication and collaboration; more efficient allocation of resources; improved access to information

By Brian L. Long, PE; June Chang, PE; and Derek St. John, PE

Waterline leaks, sewer breaks, deteriorating roads, and failing bridges might sound like typical third-world challenges, but these are issues all Americans are going to have to face in the coming years as lack of funding to support the growing population's infrastructure needs limits conventional solutions. This looming crisis of crumbling infrastructure will affect everyone.

Houston's infrastructure, like that of many other major metropolitan areas, is aging, and in many instances requires considerable restoration. As Houston developed over time, roads, waterlines, and sanitary and storm sewers were installed at a frenzied pace to meet the demands of the steadily growing suburban population. Now, more than thirty years later, many of these infrastructure components have deteriorated considerably, are operationally and/or structurally questionable, and are subsequently in need of significant rehabilitation or replacement to prevent their inevitable failure.

According to a report by the American Water Works Association (AWWA) released in May 2001, replacing the nation's aging basic water infrastructure will cost an estimated $250 billion over the next thirty years. AWWA officials believe that although the country's existing water system has been constructed in spurts since the late 1800s, distribution system piping and other water transmission components will all reach the end of their useful life at approximately the same time because of changes in materials and construction techniques over time.

Although Houston recognizes the importance of replacing its aging water infrastructure, it is not economically feasible to replace all of the existing system components simultaneously. This situation presented a complicated challenge for Houston, which was tasked with developing a method for infrastructure replacement prioritization that would maximize the benefits of the financial investment while meeting the needs of its citizens.

For many municipalities, major capital-improvement project decisions are politically motivated and prioritized to benefit those individuals or groups with the loudest voice and/or the deepest pockets. This reactive approach does little to solve the overall problem of how to replace or rehabilitate aging local infrastructure. Rather than follow a politically driven repair-and-replacement schedule, the City of Houston opted to implement a methodology that gives precedence to waterlines based on an analysis of available historic and scientific data, objective criteria, and sound engineering principles. This type of prioritization program will eventually save the city millions of dollars in restoration costs and will protect Houston's citizens and businesses from the crippling effects of broken waterlines.

This type of prioritization program will eventually save the city millions of dollars in restoration costs and will protect Houston's citizens and businesses from the crippling effects of broken waterlines.

The project objective was to provide a water system work-order data analysis for Houston, using the city's GIS-based information asset-management system to reduce future water main repairs by effectively prioritizing future replacement projects. The work consisted of analyzing existing water main work-order history; compiling work-order data into field, management, and executive-level reports; assisting in management of field-level activities and work-order database input; determining status of repairs and replacement; and confirming compliance of process and procedures. GIS helped determine problem areas and proactively plan for future improvements.

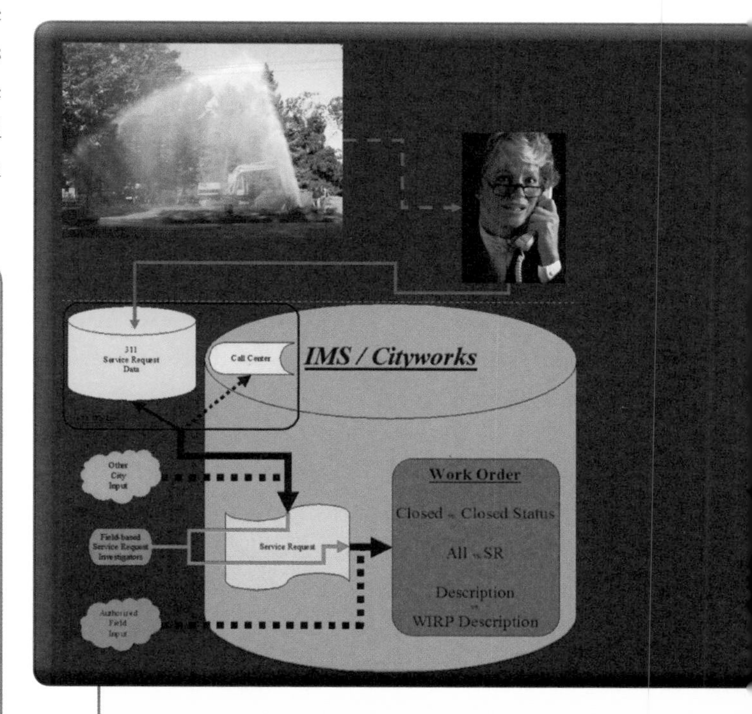

Figure 2.4 The City of Houston's Water Infrastructure Replacement Prioritization program uses GIS to prioritize which waterlines to fix first. Work-order history figures prominently in planning for future improvements.

The City of Houston's Water Infrastructure Replacement Prioritization (WIRP) program takes advantage of its investment in GIS and enterprise infrastructure asset-management system to provide the data, software, and problem-solving algorithms necessary to support the systematic prioritization of small-diameter waterline replacement projects within the city's existing system. Specifically, the WIRP selection process is based on eighteen waterline and historical water repair categories developed by city staff members, and local engineering consultant and ESRI business partner Lockwood, Andrews & Newnam Inc. The various evaluation categories are weighted to place emphasis on the more influential criteria, and are then entered into a system developed to prioritize the highest need for replacement. Ultimately, simple visual exhibits of the prioritized projects are produced through the combined use of GIS and the asset-management system, and provided to local decision makers for consideration.

Using the city's geographic information and management system (GIMS), along with rational data-populating algorithms, hundreds of thousands of GIS database records were analyzed to find approximately 456 miles of waterlines classified as critical for replacement at an estimated cost of $180 million. Although this is good data to have, it was not enough. The city could not afford to replace $180 million in waterlines that year, or even in the next five years. The WIRP project went a step further and developed a more detailed list of criteria to prioritize which waterlines would be replaced first.

A chaotic collection of required replacement projects was reduced to an orderly compilation of manageable investments. The return on investment was realized within the first year after projects were recommended on the draft CIP.

Replacement ranking criteria

Determining and appropriately weighting criteria for a system to replace small-diameter waterlines was not an easy task. The first step was to decide which characteristics are considered "substandard" for a waterline and rank it on the replacement list. After reviewing the existing water system GIS data that was reliably available, the WIRP team defined a substandard waterline according to the following criteria:

- Size—diameter less than six inches
- Material—asbestos cement, cast iron (not ductile iron), and galvanized metal
- Age—older than 40 years

Waterlines that meet these criteria have a high replacement priority because of inadequate fire protection capabilities, poor water quality, low-pressure problems, indications of redundant parallel lines, and a high probability of failure.

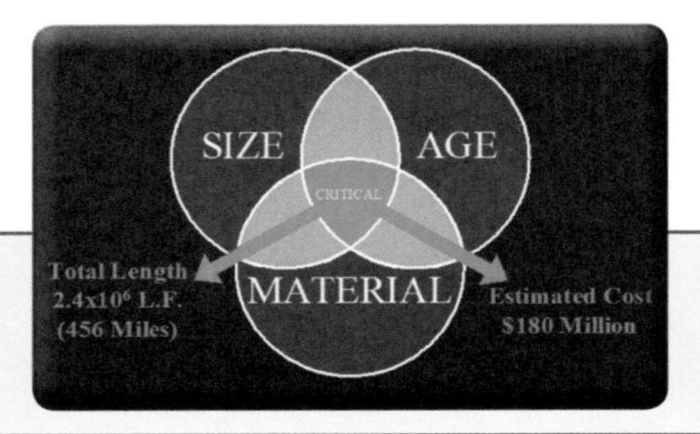

Figure 2.5 By analyzing hundreds of thousands of GIS database records, the City of Houston found 456 miles of waterlines deemed most critical for repairs, at a projected cost of $180 million.

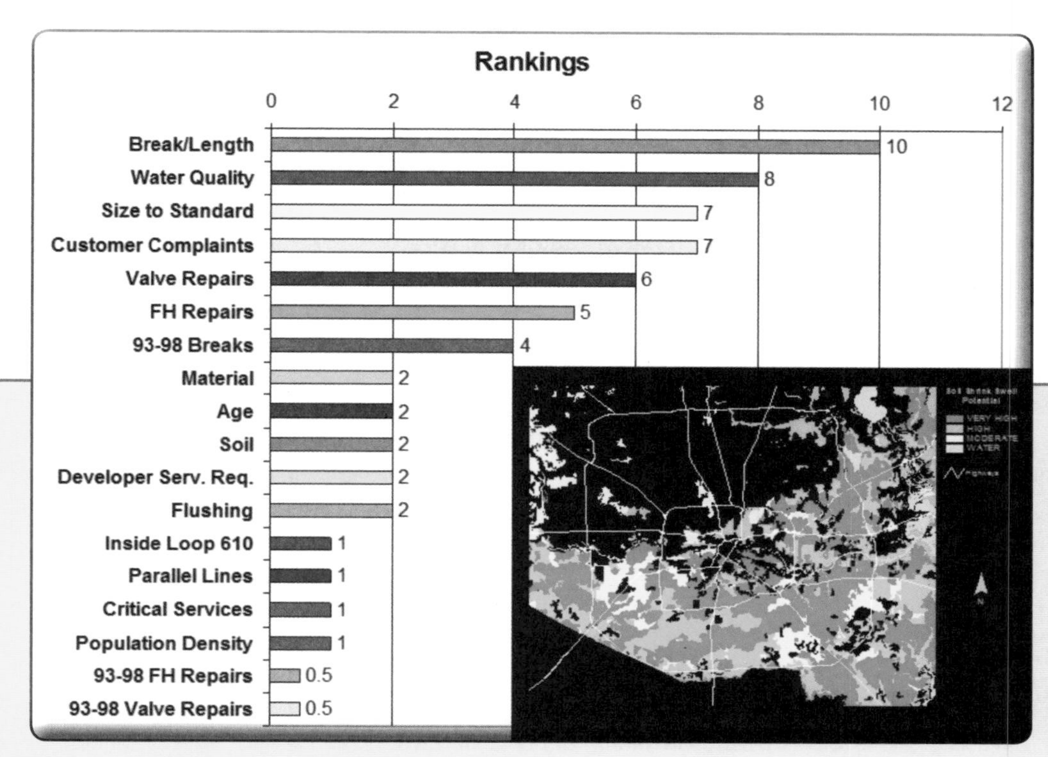

Figure 2.6 Because the city could not afford $180 million worth of repairs all at once, it relied on eighteen weighted factors to determine its greatest infrastructure needs.

The city ultimately settled on eighteen weighted factors to determine replacement priority. The factors and their relative weights were determined after many consensus-building meetings. Most of the data used for each factor came from work orders and asset inventories stored in the city's integrated infrastructure asset management and GIS systems. The selected criteria are as follows:

1. **Waterline breaks (breaks)**—Ranks map grid blocks according to the number of breaks within the block that occurred from July 1999 to 2002 divided by the total length of waterlines (i.e., normalized by waterline length) within the block that are less than or equal to twelve inches
2. **Water quality**—Ranks map grid blocks according to the number of water-quality complaints within the block divided by the total length of waterlines (i.e., normalized by waterline length) within the block that are less than or equal to twelve inches
3. **Size to standard**—Ranks map grid blocks according to the length of lines less than six inches within the block divided by the total length of waterlines (i.e., normalized by waterline length) within the block that are less than or equal to twelve inches
4. **Valve repair**—Ranks map grid blocks according to the number of valve repair work orders within the block that occurred

from 2000 to 2002 divided by the total length of waterlines (i.e., normalized by waterline length) within the block that are less than or equal to twelve inches

5. **Fire hydrant repairs**—Ranks map grid blocks according to the number of fire hydrant repair work orders within the block that occurred from 2000 to 2002 divided by the total length of waterlines (i.e., normalized by waterline length) within the block that are less than or equal to twelve inches
6. **Customer complaints**—Ranks map grid blocks according to the number of customer complaints on twelve inches of waterline that are within the block
7. **1993–98 waterline breaks**—Ranks map grid blocks according to the number of breaks within the block that occurred from 1993 to 1998 divided by the total length of the waterlines (i.e., normalized by waterline length) within the block that are less than or equal to twelve inches
8. **Material**—Ranks map grid blocks according to the length of asbestos cement and cast iron lines within the block divided by the total length of waterlines (i.e., normalized by waterline length) within the block that are less than or equal to twelve inches
9. **Age**—Ranks map grid blocks according to the following age increments: <10=1, 10-20=2, 20-30=3, 30-40=4, 40-50=5, >50=6. The sum of the age increments within the block is divided by the total length of waterlines (i.e., normalized by waterline length) within the block that are less than or equal to twelve inches.
10. **Soil type**—Ranks map grid blocks by assigning each line segment a soil type from 1 to 4 with 4 the worst case. The sum of the assigned soil types numbers for each block is divided by the total length of waterlines (i.e., normalized by waterline length) within the block that are less than or equal to twelve inches.
11. **Developer service requests**—Ranks map grid blocks according to the number of developer service requests within the block.
12. **Flushing frequency**—Ranks map grid blocks according to the number of line-flushing procedures within the block divided by the total length of waterlines (i.e., normalized by waterline length) within the block that are less than or equal to twelve inches.
13. **Location of waterline inside Loop 610**—Ranks map grid blocks based on percentage of 12 inches of waterline within the block that is also within Loop 610.
14. **Two-inch parallel lines**—Ranks map grid blocks according to the length of two-inch parallel lines within the block divided by the total length of waterlines (i.e., normalized by waterline length) within the block that are less than or equal to twelve inches.
15. **Dialysis centers**—Ranks map grid blocks according to the number of dialysis centers within the block.
16. **Population density**—Ranks map grid blocks according to the total population within the block based on the Census 2000 population count.

17. **1993–98 fire hydrant repairs**—Ranks map grid blocks according to the number of fire hydrant repairs within the block that occurred in 1993 to 1998 divided by the total length of waterlines (i.e., normalized by waterline length) within the block that are less than or equal to twelve inches.
18. **1993–98 valve repairs**—Ranks map grid blocks according to the number of valve repairs within the block that occurred in 1993 to 1998 divided by the total length of waterlines (i.e., normalized by waterline length) within the block that are less than or equal to twelve inches.

Prioritizing replacement

The refined prioritization method used map grids overlaid on the greater Houston area. Each grid (approximately one-half square mile) encompassed an area substantial enough to support most subdivisions yet still small enough to be representative of the required project size.

After the composite values of the factors were totaled for each map grid, the grids were ranked and their relative priority determined. The fifty grids with the highest rankings were then considered for the city's five-year capital improvement plan (CIP), with the projects in the highest ranked grids recommended for funding in the first year. At least one specific replacement project was recommended for each CIP-recommended grid on the chart. Not every highly ranked grid on the chart was recommended as a priority for a replacement project for CIP consideration. This shows the flexibility of the WIRP project selection system in prioritizing the city's greatest water infrastructure needs. Those grids highly ranked but not chosen were removed from CIP consideration either because of the grid's preexisting inclusion in the CIP or because of the overlap of a subdivision onto two grids, thereby forcing the selection of one grid over another for the representative project.

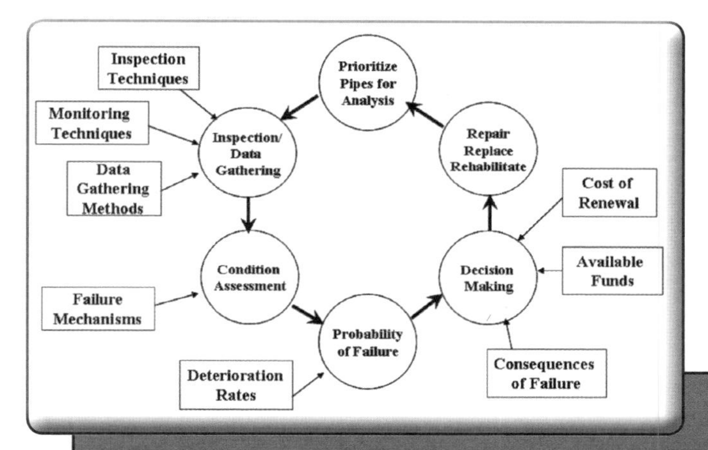

Figure 2.7 Through the use of GIS via its WIRP program, the City of Houston can effectively manage millions of dollars worth of repairs even in tight budget times.

As a result of the WIRP CIP prioritization analysis, the city was able to reduce the estimated $180 million replacement cost to $59 million spread over five years for thirty-one projects in the areas of the city most in need of waterline replacements. A chaotic collection of required replacement projects was reduced to an orderly compilation of manageable investments. The return on investment was realized within the first year after projects were recommended on the draft CIP. As the decision makers, directors, and elected officials reviewed the projects, the WIRP data provided the utility planning selection with objective, fact-based justification for the recommended projects. With the GIS-based WIRP reports as supporting documents, the recommendations were accepted with confidence and efficiency.

Through the WIRP, the City of Houston has heeded the call of its citizens to manage the city's infrastructure needs within the limited budget that today's economy dictates. Because Houston is faced with reduced revenues and a crumbling infrastructure, its leaders have turned off the road leading to inefficient use of limited funds and are boldly taking the path of fiscal responsibility. In Houston, the true value of GIS and infrastructure asset management systems is realized as tools that increase efficiency and reduce the costs of government.

Houston has made dramatic strides in the efficient repair and replacement of its aging water infrastructure in recent years, which will continue. While this article focused on water infrastructure solutions, the methodologies and applications developed are adaptable to other infrastructure as well. Houston's city staff hopes to apply the infrastructure replacement prioritization process to sewerage, drainage, and road infrastructure in the future.

As a result of the WIRP CIP prioritization analysis, the city was able to reduce the estimated $180 million replacement cost to $59 million spread over five years for thirty-one separate projects in the areas of the city most in need of waterline replacements.

Assessing environmental vulnerability of Boston Harbor

CHAPTER: Defending a decision/reaching a compromise
ORGANIZATION: U.S. National Park Service
LOCATION: Jamaica Plain, Massachusetts
CONTACT: Jennifer Bender, PhD, geographer
ferre@alum.bu.edu
PROJECT: Environmental vulnerability assessment
SOFTWARE: ArcView
ROI: Increased efficiency, accuracy, and productivity

By Jennifer Bender

The Boston Harbor Islands National Recreation Area (BHINRA) was created in 1996 by congressional act (Public Law 104-333). In establishing the Boston Harbor Islands as a unit of the National Park Service (NPS), Congress created a governance structure that was unique in several ways.

First, BHINRA is a public-private partnership. Operating under the auspices of the U.S. Interior Department, the NPS has been directed to operate as a nonlandowning participant in the partnership. This invests the NPS with the responsibility, but not the authority, to make decisions directly related to the congressional mandate to achieve ecological, educational, recreational, and economic goals. Authority to determine policy and make decisions rests instead with the partnership, a thirteen-member body of representatives from public, private, and nongovernmental organizations.

The second unique feature is the diverse ownership of the islands. BHINRA includes thirty-four islands in Boston Harbor that are owned by a variety of federal, state, nonprofit, and private entities.

Congress did not seek to transfer all ownership to the federal government, but instead created a legal and administrative framework through the National Recreation Area (NRA) designation that would protect the islands through decision making that took into account the overarching goals.

Figure 2.8 The Boston Harbor Islands National Recreation Area, which comprises thirty-four islands, is America's newest national park.

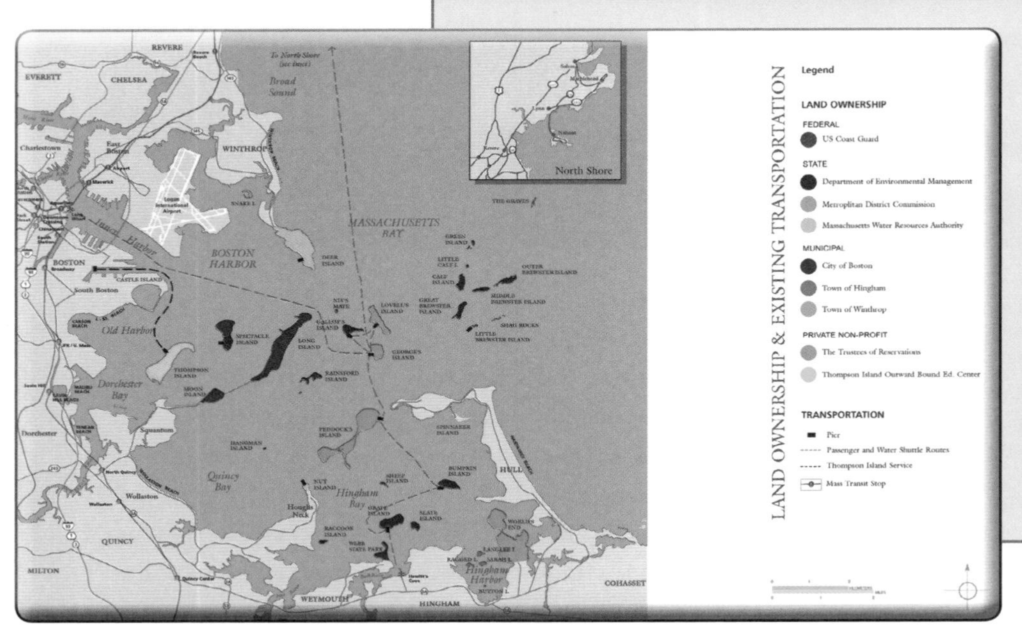

Figure 2.9 Map of the Boston Harbor Islands National Recreation Area shows how ownership of the islands is split among federal, state, nonprofit, and private entities.

Third, funding for the BHINRA is not solely a federal responsibility. Congress decided that the National Recreation Area would be funded through a partnership arrangement in which the federal government would provide operating funds on a matching basis, with local community entities (state, nonprofit, corporate, and private) providing the remaining support.

These governance features make the administration of the BHINRA a complicated undertaking. Each feature generates pressures that have influenced short-term actions, overall priorities set forth in the enabling legislation, and long-term strategies to develop the islands. These factors also make BHINRA an exceptionally interesting case study and a potential twenty-first-century model for the nation. Many of the factors that Congress weighed in creating BHINRA affect other national parks as well. Mixed ownership, public-private partnerships, and diverse sources of funding are often considered in the design of strategies to protect and preserve scarce natural resources. Collaborative models are of great interest to public, private, and nongovernmental decision makers. BHINRA will represent an increasingly important precedent as it matures and lessons are learned about the decision process and the outcomes of various choices.

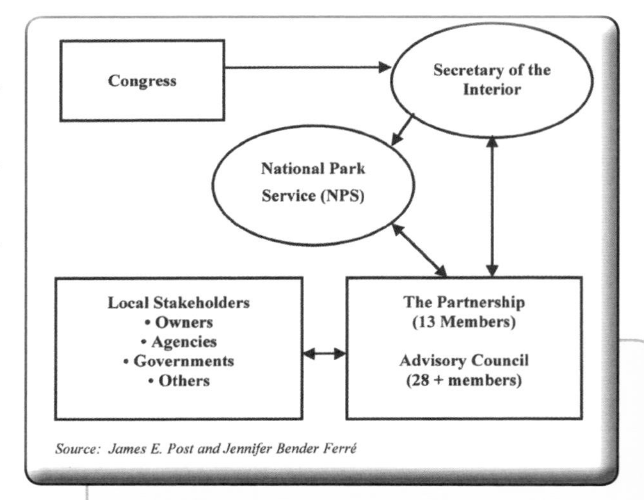

Figure 2.10 Chart of the BHINRA governance structure shows how responsibilities and support for the national recreation area are divvied up.

Project background

An overview of environmental problems and governance structures in BHINRA revealed specific management problems and challenges. Since Congress created BHINRA as America's newest national park, planning has been fragmented, with no comprehensive program to establish sustainable-use patterns. The founding legislation defines a new model regarding landownership and sets a new standard for park funding.

Sustainable-use patterns use an integrated approach to consider the entire ecosystem. Ecosystem management is a necessary piece in the governance of decision making, and linking marine governance to the paradigm of sustainable development based on an ecosystem approach will require the creation of new tools to support complex decision processes.

Subsequently, a decision-support tool was developed to assess environmental vulnerability to boating in BHINRA. A modified "weight of evidence" method was used to elicit information from a panel of scientific experts, and the information was entered into a GIS-based decision-support tool.

The process was initiated with a workshop that elicited scientific expertise relevant to decisions around increased stressors such as boating in BHINRA. The outcome of the workshop was the definition of ecological endpoint goals (EEGs) and measures of effects (MOEs) to then be applied to a GIS-based map. The EEGs were defined as the explicit expressions of the environmental elements to be protected, while the MOE is statistical or arithmetic summaries of the chosen EEG. The MOE is the basis for structuring the analysis phase and can serve as actual measurements.

GIS Initial list of ecological endpoint goals (EEG)
Maintain intertidal migration
Protect migratory seabirds
Prevent invasive species
Maintain submerged intertidal communities
Maintain seal communities
Protect against smothering
Improve water quality
Protect air quality
Prevent noise pollution
Protect intertidal and subtidal near-dock environments
Protect migratory fish
Maintain recreational fishing
Protect nesting water birds
Prevent land disturbance of birds
Protect winter waterfowl
Restore coastal features, dunes—landward side of intertidal environment
Maintain shore-bird habitat
Restore marshes
Restore eelgrass
Protect near-shore substrate
Protect habitat of migratory and endemic species
Maintain some inner islands as pristine
Replant native vegetation

Table 2.2 A scientific workshop defined EEGs as the environmental elements to be protected as a result of increased boating in BHINRA.

Ecological Endpoint Goal					
		Potential for Impacts from Boat Activity			
		1.	**2.**	**3.**	**4.**
		Ecologically Important Habitat Restoration	**Maintain and Improve Water Quality**	**Protect Aquatic and Coastal Habitats and Species Indicators**	**Maintain and Improve Air Quality Indicators**
Measures of Effect	**A**	Salt-marsh and eelgrass area distribution	Water Turbidity (TSS)	Change in Substrate Type	Photochemical smog
	B	Area and distribution of shellfish beds	Dissolved Oxygen	Wake Induced Erosion. Ferry routes slope greater than drumlins can be vulnerable if wave because mounds of sediment.	Diesel Particulates
	C	Shoreline Nesting Habitat (tern productivity)	Nutrient Loading	Nesting Water bird Productivity	CO
	D	Decline in Invasives Associated with degraded habitats (phragmites)	Pathogen Indicators	Change in Seal Haul Out Areas	Noise level db limits
	E		Petroleum Hydrocarbons	Litter Covered Area	
	F		Synthetic Pollutants	Area Impacted by Invasive Species	
	G		Species Indicators	Changes in Inter-tidal Communities: Rocky	
	H		Floatables	Changes in Inter-tidal Communities: Soft	
	I			Fish Populations Distribution Migration	
	J			Migratory Shorebird Population Distribution	
	K			Wintering Waterfowl Number Diversity Distribution	

GIS was an ideal tool to support a group-based decision methodology, because it permitted visualization of the complex ecological data as well as the EEG generated from the workshop.

Figure 2.11 The workshop defined ecological endpoint goals and measures to evaluate the effects or benefits associated with each ecological endpoint goal.

GIS was an ideal tool to support a group-based decision methodology, because it permitted visualization of the complex ecological data as well as the EEG generated from the workshop. The goals were integrated into a final map divided into grid cells that indicated the number of critical MOE present in each cell, and the indicators were weighted based on the results of the workshop. The results were displayed in several layered maps of measures of environmental concerns and their associated goals. These maps would enable decision makers to visualize and prioritize impacts of potential increased boating activities. The result was an intuitively appealing comprehensive and interactive tool that aids decision makers.

Science and policy

The role of science is not defined in either the legislation or the partnership documents governing BHINRA. Instead, it seems to be assumed that scientific issues will be raised as they occur in the context of specific plans, projects, and activities. The advisory council contains no scientific committee to identify issues or to design protocols for investigating scientific questions: It appears that scientific expertise is brought in on a "stand-by" or "as needed" basis.

In this project, we consider how science can be brought to bear in systematically on policymaking for BHINRA's future. Political systems have been superimposed on natural systems, with many stakeholders woven into the decision-making fabric. This report addresses the interaction between people and the environment, and the governance issues that arise when the physical environment and political environment clash.

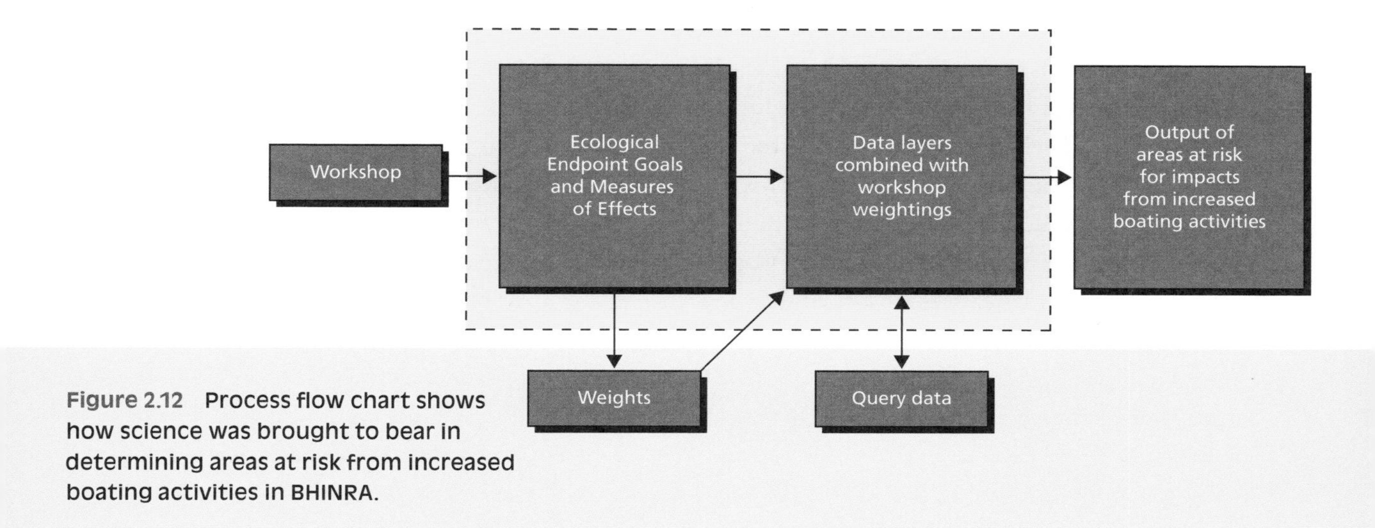

Figure 2.12 **Process flow chart shows how science was brought to bear in determining areas at risk from increased boating activities in BHINRA.**

The central question is, How, within a defined governance system, can the public capitalize on scientific knowledge and expertise to better achieve public goals and objectives? It is clear that the management of complex ecosystems requires knowledge and insights from the natural and social sciences to formulate and implement appropriate management strategies. One of the major challenges to the effective management of coastal resources, and BHINRA in particular, involves the interface between science and policy. There is a tenuous balance to be struck between a scientific community that answers operational research questions for public policy makers, and policy makers who ask the proper questions of the scientific community.

Boston Harbor ecosystem

Boston Harbor and the thirty-four islands in BHINRA are part of Massachusetts Bay. The harbor islands are contained within fifty square miles, including land and water. (It is noteworthy that the NRA does not include the water.) The harbor is flanked by many communities, including Chelsea, Charlestown, Dorchester, downtown Boston, Fort Point Channel/New Seaport District, East Boston, Hingham, Hull, Quincy, Revere, South Boston, Weymouth, and Winthrop. The islands range in size from one-quarter acre (Hangman Island) to 214 acres (Long Island). Together, the thirty-four islands make up 1,200 acres of land.

The majority of the islands are drumlins (elongated, unstratified glacial till deposited and molded below an ice sheet), and a few of the islands are outcrops of bedrock. The vegetation and landscape, which has been mostly altered over the centuries, provides varying habitats for plants and animals. The islands provide special microenvironments that offer a unique combination of natural resources in an area surrounded and bordered by the ocean; the water, marshes, and open areas are host to a variety of birds, mammals, finfish, and shellfish.

An important scientific dimension of the Boston Harbor Islands is their relationship to other marine ecosystems. The islands are contained within Massachusetts Bay, which, in turn, is part of the Gulf of Maine. Massachusetts Bay comprises approximately 1,400 square miles bounded on the north by Cape Ann (Gloucester) and on the south by Cape Cod (Provincetown). Its easternmost boundary is Stellwagen Bank National Marine Sanctuary, which encompasses eight hundred square miles. The Gulf of Maine watershed extends from Quebec, Nova Scotia, and the Bay of Fundy in the northeast to Boston and Cape Cod in the southwest. The total land area of this watershed is 69,115 miles.

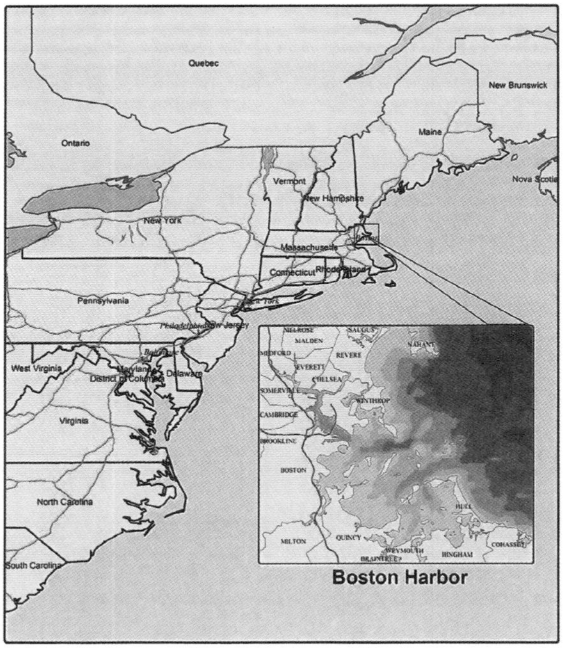

Location of Boston Harbor

This map shows the location of Boston Harbor in relation to the East Coast. The data used for the map is from ESRI and the Massachusetts office of geographic information (MassGIS).

Figure 2.13 Boston Harbor is in Massachusetts Bay, which is part of the Gulf of Maine.

The Gulf of Maine is the largest semienclosed sea shelf bordering the continental United States and features almost every conceivable use of the marine environment. The water and islands are interdependent and must be considered in the development of an integrated management plan. The intertidal, coastal, ocean-adjacent, and terrestrial island areas represent a complex and dynamic environment in which chemical, geological, biological, meteorological, and estuarine processes take place. The islands contain a fairly uniform marine environment with consistent scientific features, but the coastal area is characterized by a variety of forms: rocky shores, sandy beaches, estuaries, lagoons, intertidal flats, wetlands, and islands. These habitats are interlinked and should be considered a unified system. This underlying sea-land interaction is at the center of the ecological learning opportunity in BHINRA.

Methodology

The weight-of-evidence approach involves the development of a framework to capture the relative importance assigned to various factors by panels of experts. A framework was used because decisions concerning management alternatives need a transparent means of incorporating expert opinion into the decision-making process. A modified and adapted weight-of-evidence approach has been introduced to help bridge comparisons of economic, environmental, and societal values.

Under this type of analysis, all available evidence is considered, and a conclusion is reached based on the amount and quality of evidence supporting alternative outcomes. MOEs are related to EEGs to evaluate whether a significant risk of harm is posed to the environment. The approach is initiated at the problem-formulation stage, and results are integrated at the risk-characterization stage. The analysis includes the following key concepts:

- EEG: an explicit expression of the environmental value to be protected
- MOE: a measurable ecological characteristic of an agent (a physical, chemical, biological entity that can induce an adverse or beneficial response) that is used to quantify exposure of the ecological endpoint

EEG1 was determined to be ecologically important habitat restoration. And the four primary MOEs with this goal were defined as

- MOE A: Salt marsh and eelgrass area distribution
- MOE B: Shoreline nesting habitat (tern productivity)
- MOE C: Area and distribution of shellfish beds
- MOE D: Decline in invasives associated with degraded habitats

Measures	Weighting Score (1-5)	Relationship Between Measure of Effect & Ecological Goal	Degree of Association Between Boat Activity and Measure of Effect	Magnitude	Evidence of Harm (yes/no/ undetermined)
MOE A1 Salt marsh and eelgrass area distribution	3.9	4.1	2.1	high	undetermined
MOE B1 Shoreline Nesting Habitat (tern productivity)	3.6	3.9	2.2	high	yes
MOE C1 Area and Distribution of Shellfish Beds	2.8	3.4	2.6	high	yes
MOE D1 Decline in Invasives Associates w/Degraded Habitats	2.9	3.2	1.3	low	no

Figure 2.14 A score sheet tallies the evidence of harm and magnitude for Endpoint Goal 1, ecologically important habitat restoration.

Scientific workshop

An all-day workshop was held September 23, 2002, at the New England Aquarium in Boston to develop a set of measures that could be used in the process of environmental decision making regarding boating activities.

To meet the criteria of the enabling legislation, visitation to the islands had to be doubled. This could have consequences including wake effects, pollution, and noise. Two types of measures had to be identified: EEG and MOE related to these endpoints.

Workshop results established EEG and their associated MOE, determined the magnitude of response in the MOE, and provided the ability to evaluate concurrence between MOE and EEG.

In addition, the results enabled researchers to map and define areas identified in the workshop as areas that were potentially sensitive pertaining to the different MOE established in the workshop.

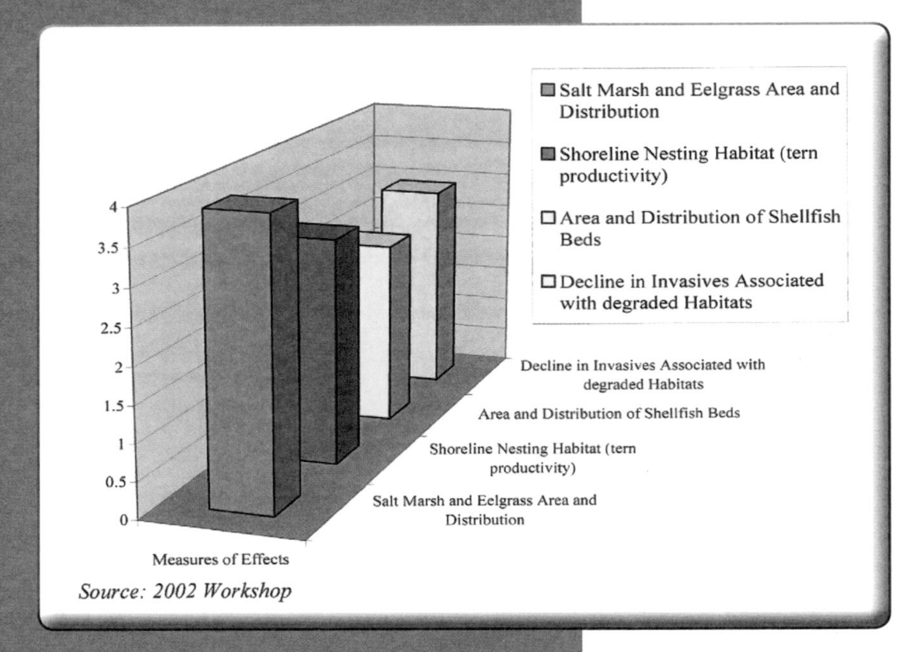

Figure 2.15 Chart shows the four primary MOEs for the goal of habitat restoration.

The classification scheme consisted of layers defined by the chosen measures from the workshop, including turbidity, shellfish beds, salt marsh and eelgrass beds, multiple bird layers, and two intertidal layers. All of these layers were built on color orthodigital photos downloaded from the Massachusetts Geographic Information System (MassGIS), with an overlay of a bathymetry layer from the National Oceanic and Atmospheric Administration.

Areas of concern were identified related to increased boat traffic by defining the EEG and their associated MOE. In order to evaluate whether there is significant risk of harm, the weight of opinions was applied to a series of map layers indicating potentially vulnerable areas on the map.

Classification criteria and patterns of measures

A hierarchical classification criterion was influenced primarily by factors arising out of the 2002 workshop. This process included spatial analysis based on the weight-of-evidence criteria developed by workshop participants. The maps reflected the relative importance of the MOE, based on the weighted averages derived from the input of workshop participants.

Of the nine layers mapped, four layers were analyzed in detail to illustrate the interconnectedness among measures. These measures underscored the complexity of the system, because they were dependent on one another. They included turbidity, shoreline nesting habitat (tern productivity), salt marsh and eelgrass, and intertidal soft. The GIS maps used a simple algorithm to quantify grid cells representing locations throughout the Boston Harbor Islands Area. Different layers and grids were assigned weights based on criteria established in the workshop. The results were displayed as a raster or grid map.

Workshop attendees defined the goals and prioritized the EEG according to stakeholder input and scientific opinion. This method enabled policy makers and stakeholders to view the system in its entirety.

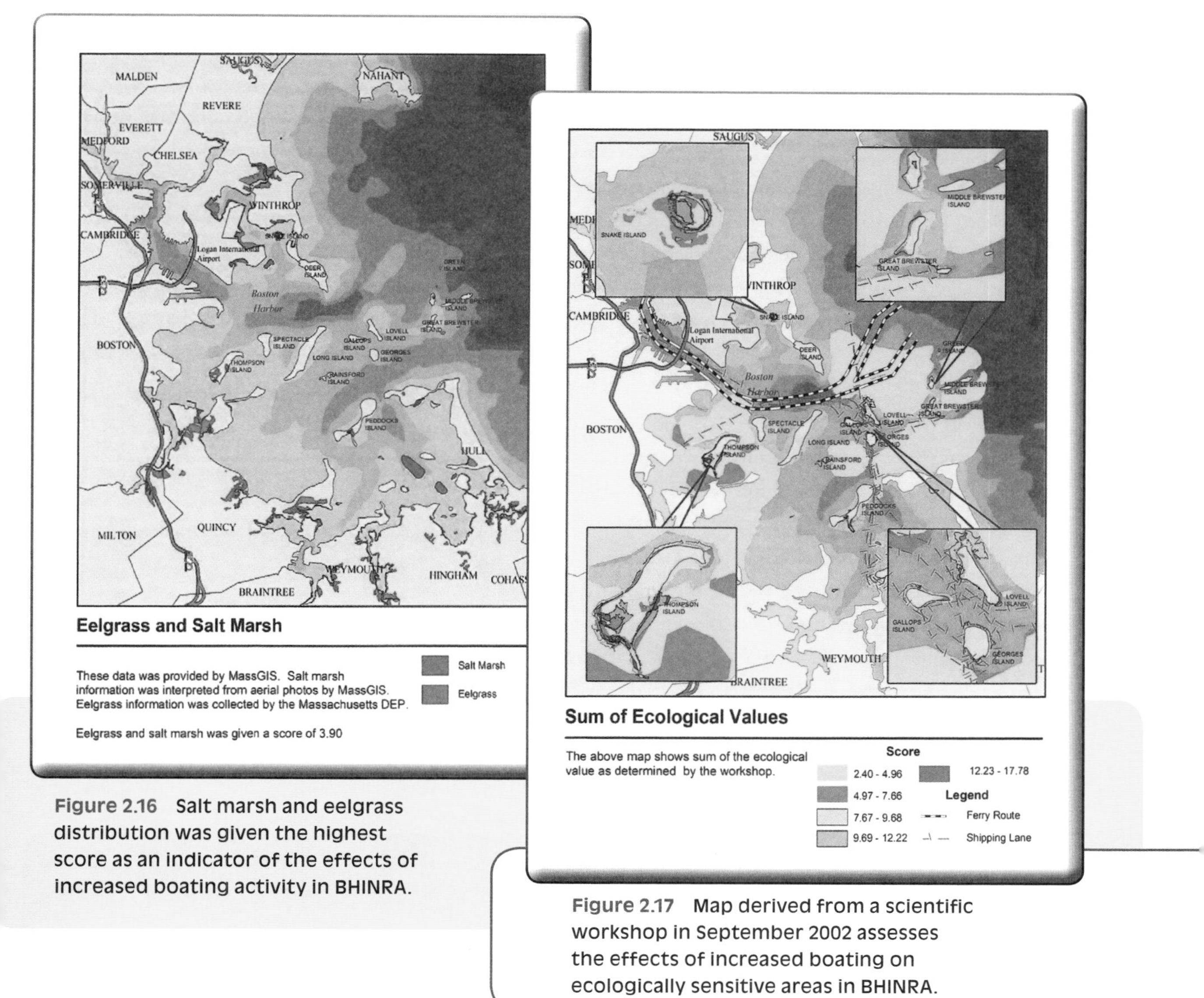

Figure 2.16 Salt marsh and eelgrass distribution was given the highest score as an indicator of the effects of increased boating activity in BHINRA.

Figure 2.17 Map derived from a scientific workshop in September 2002 assesses the effects of increased boating on ecologically sensitive areas in BHINRA.

Data layers were overlaid characterizing each of the selected measures. The more layers one could apply, the broader the understanding of the system. There is little historical precedent in managing whole ecosystems, and this process provides a macroview using detailed comprehensive information. In this case, the exercise was to capture one issue—boating activity—as a stressor on a chosen set of goals and measures in the whole island system, and then quantify it scientifically, apply it spatially, and juxtapose the interrelationship. The evidence presented demonstrated that potential disturbance exists, and it helped inform policy decisions.

The GIS model produced maps identifying the spatial patterns of ecological vulnerability to boat traffic in the islands and surrounding waters. The maps were an intuitively appealing, comprehensive, and interactive tool that helped decision makers and managers ultimately choose acceptable boat routes while defining "no go" areas for boats at the same time.

Return on investment

Coastal data has several dimensions inherent in its structure, and developing GIS technology has helped in visualizing the entire picture of a coastal ecology. The development of a methodology to elicit scientific expertise for incorporation into a visual format using GIS makes it easier for decision makers to use scientific opinion that has been evaluated and displayed in a prioritized and visually conclusive format. It does not pit one discipline against another but uses a consensual output of higher priority and distinction; the decision makers receive an overview of valued choices. The project shows how GIS can bridge the gaps among scientists, policy makers, and a multitude of stakeholders by collecting new data, incorporating sound science, and adapting a new methodology that can shape data into a policy-relevant mechanism. It can provide decision makers with an overview of valued choices that can validate their decisions.

The maps were an intuitively appealing, comprehensive, and interactive tool that helped decision makers and managers ultimately choose acceptable boat routes while defining "no go" areas for boats at the same time.

Using GIS in tribal negotiations

CHAPTER: Defending a decision/reaching a compromise
ORGANIZATION: Oneida Nation
LOCATION: Oneida, Wisconsin
CONTACT: Celene Elm, GIS/indigenous planning director celm@oneidanation.org
PROJECT: Taxation
SOFTWARE: ArcGIS
ROI: Increased efficiency, accuracy, and productivity; improved access to information

By Celene Elm

All tribal nations negotiate with governments, and the effectiveness of these negotiations can have a significant impact on both the environment and the people. The Oneida tribe anticipates the annual negotiation of its gaming compacts, but sometimes unanticipated opposition brings them to the bartering table. With GIS, the tribal council can quickly and effectively prepare for these sessions.

Recently, the Oneida Nation found itself in the position of having to defend its tax-exempt status. On the local level, the tribal nation has been entering into cooperative agreements with surrounding governments to contract services. The Oneida Reservation exists within two counties and four townships, and all jurisdictions negotiate with the Oneida Nation for services.

A major point of contention often focuses on the nation's lands in trust. Taxation status serves as the criteria for arbitrating what the tribe is expected to pay for services, with the counties and townships both claiming that the area of land the nation holds in trust is so vast that it impacts both the county and municipal tax base. The Oneida have disputed these claims and with GIS technology were able to disprove them.

By examining the workbook assessment, the tribe's GIS team mapped out the area according to tax classifications. The first step was to access the database for files of roads, reservation boundaries, ownership data, and parcel maps. Next, the team employed the GIS query tools to show themes: tribal-fee property, tribal trust, and individual-fee property; and roads, road names, parcels, parcel identifications, and individual-fee property. Then it referenced the workbook assessments and matched this information with parcel numbers to determine the acreage of the tax classification. The last step was to complete the cleanup and edits. The final product was a map depicting the ratio of land by tax classifications.

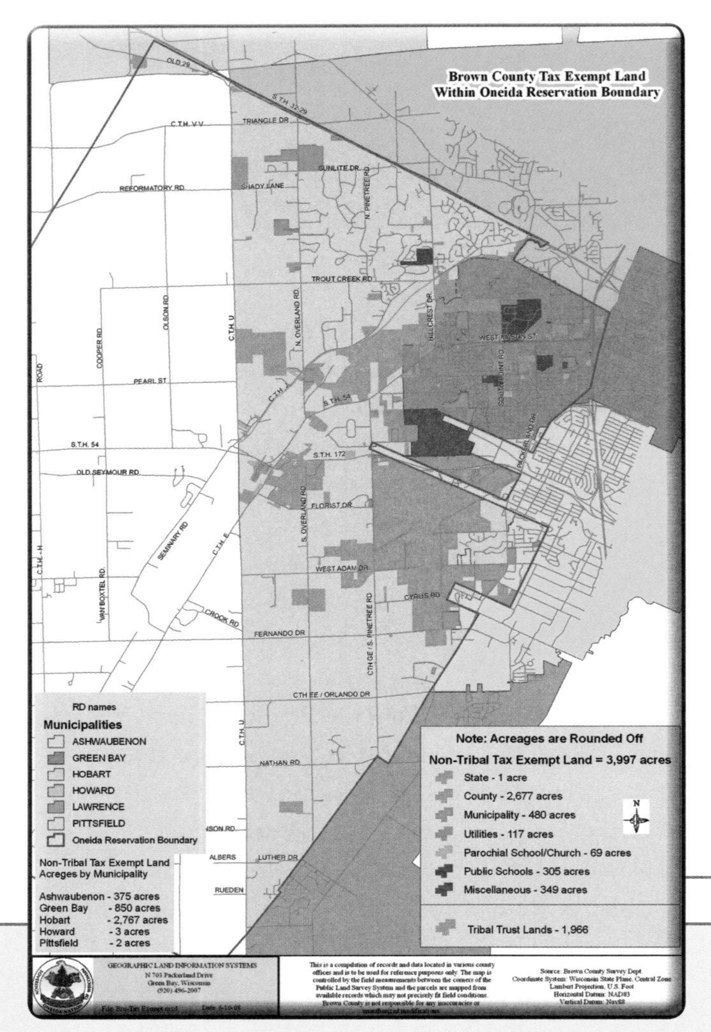

Figure 2.18 Map shows that the Oneida Nation holds less land in tax-exempt status than surrounding townships do.

Within just eight hours, the GIS team produced results that supported the tribe's position. The maps easily illustrated the reality—townships held more land in tax-exempt status than did the Oneida Nation.

Without the strength of GIS analytical tools, tax classifications would certainly have been misrepresented. This would have resulted in the tribe having to pay too much when contracting service agreements. Bringing GIS to the table offers credible mapping that is based on real data, which significantly impacts decision making and negotiations.

Without the strength of GIS analytical tools, tax classifications would certainly have been misrepresented.

Exercise in decision support: Defending a decision

Where to build a new fire station

Government decisions are often politicized, and issues can be decided for reasons other than merit. Elected officials and government professionals can be swayed by a vocal minority or unduly influenced by anecdotal information. By dispassionately analyzing and presenting the facts of a situation, GIS technology can diffuse tensions and provide a defensible basis for better decisions.

In the following scenario, the city council of a rapidly growing community must decide where to locate a new fire station. The two sites that have been suggested are both on land already owned by the city. The Fifth Street site is near the city center while the Alto Dinero site is on the outskirts of the city.

Residents of Alto Dinero, an affluent section of the city, loudly advocated for locating the new fire station near their homes. Council members, who also lived in this neighborhood, felt comfortable with the decision to locate the station in Alto Dinero, because it would ensure that the area's expensive homes would likely be saved in the event of a fire. In addition, there had been little opposition from the community. Would you make the same decision as the council?

The accompanying map shows the proposed locations along with the existing stations in relation to population density.

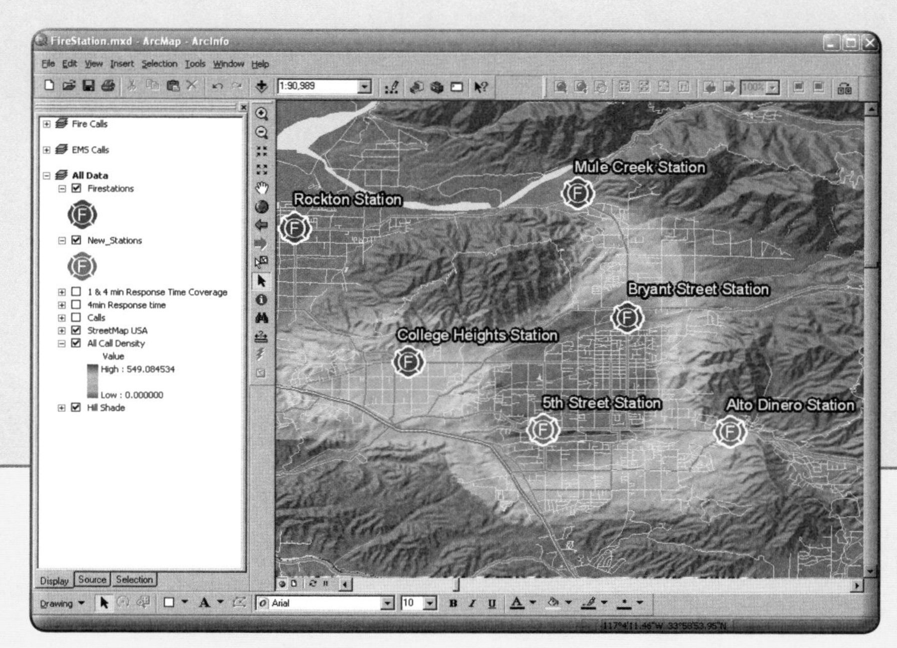

Figure 2.19 Map shows proposed versus existing fire stations in relation to population density.

New concerns, new data

Although the fire chief understood why the city council wanted to base its decision on the input from citizens, he had other concerns. He must adhere to strict National Fire Protection Association standards of coverage that require fire departments to respond to fires within four minutes.

GIS staff analyzed land-use and historical calls-for-service data to determine where to build the new station. Using the network analysis capabilities of GIS software, staff members could model response times from potential locations throughout the city's street network. This analysis demonstrated that locating the fire station in the upscale neighborhood would leave much of the populated city center without coverage.

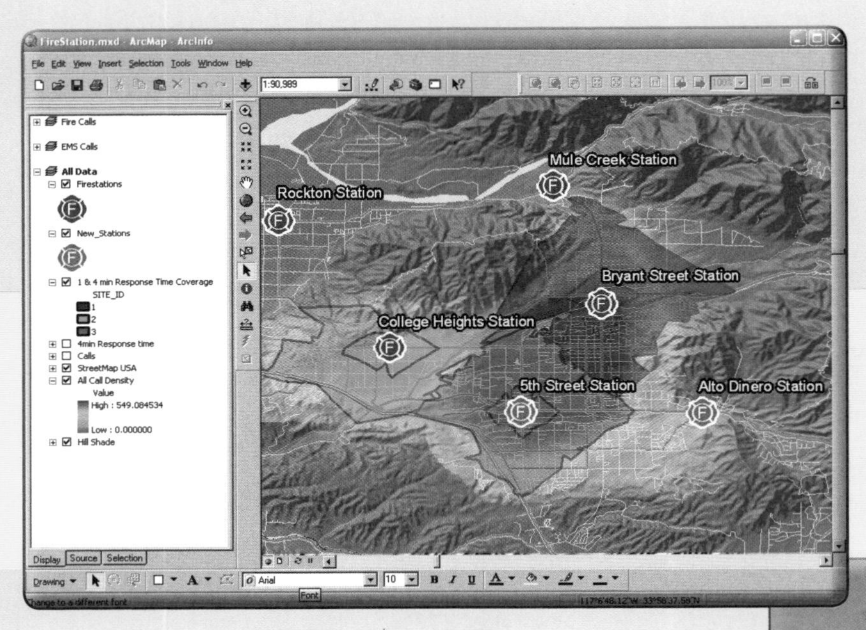

Figure 2.20 Using network analysis tools in ArcGIS, GIS staff mapped two- and five-minute coverage areas, demonstrating that the Fifth Street station provided coverage to more people.

Looking at population density and response times, staff members developed an alternate site that would provide adequate coverage for the densely populated city center. GIS analysis furnished the city council with the basis for making a better decision—one based on facts, not just emotion, and one that benefited more residents.

Figure 2.21 City fire departments are required to meet the national standard of responding to a fire emergency within four minutes.

Exercise in decision support: Reaching a compromise

Where to build—or not build—an adult business

When a city needs to develop an adult-business ordinance, how can GIS help determine where the city can permit that kind of controversial land use? The basic premise of land-use zoning is to protect property values and prevent new development from harming existing residents and businesses. However, the regulation of land use by government is often contentious. After all, as the argument goes, how can government tell private-property owners what they can or cannot do with their land?

Suppose the property owner of a vacant corner lot across the street from the local high school wants to open a 24-hour convenience store on the site. As long as the parcel is zoned for commercial use and the city cannot find a negative impact on surrounding properties resulting from that use, the store will probably be built. But what if, instead of a convenience store, the property owner wants to open an adult bookstore?

Residents typically oppose business or land development that they feel cuts into the fabric of society or infringes on their sense of decency or well-being. However, an adult business's right to operate is often protected by law. This means that cities must somehow designate where adult businesses can and cannot operate within their jurisdiction.

Officials must also consider surrounding property owners who feel that allowing an adult business to operate nearby will negatively impact the quality of their lives and businesses and reduce their property values. It is a land-use policy balancing act that can have a ripple effect of consequences in the community.

The GIS difference

When city officials are put in the awkward position of finding a suitable location for a not-so-respectable business, they need tools to help them step through a logical, repeatable process and ultimately establish an enforceable ordinance. Many of the components of an adult-business ordinance have spatial or geographic elements, such as distance, proximity, and adjacency. GIS enables stakeholders to visualize and analyze those elements, which improves the viability of the ordinance.

When a city needs to develop an adult-business ordinance, how can GIS help determine where the city can permit that kind of controversial land use?

A city might first determine which land-use zones absolutely do not permit adult businesses—for instance, open space, public facilities, and high-density residential areas. Then, land uses sensitive to adult-business activities such as medium- and low-density residential areas could be identified.

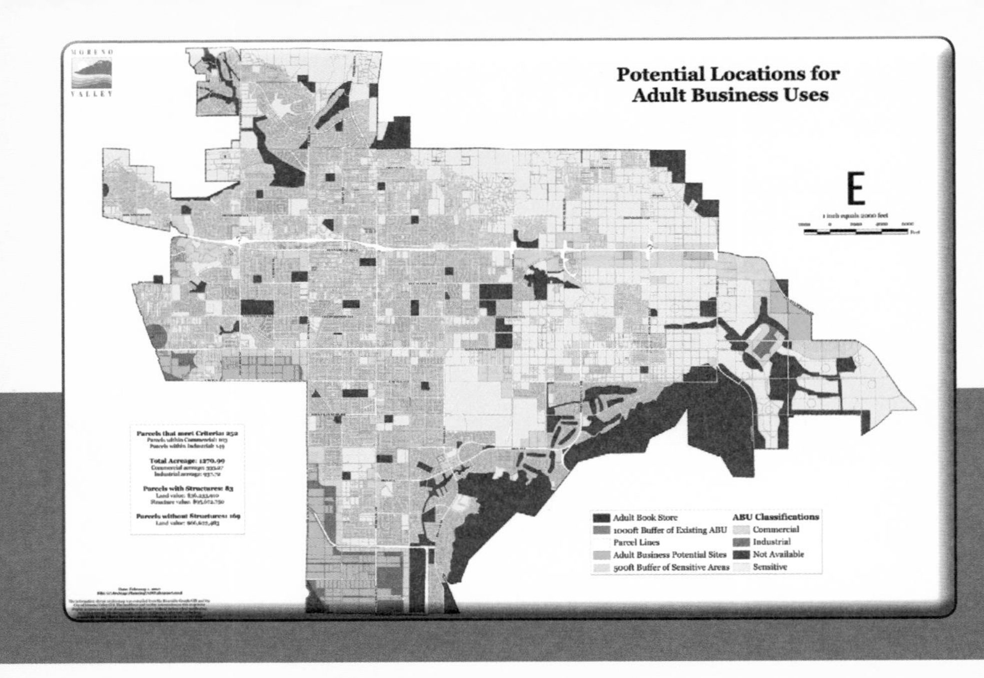

Figure 2.22 With GIS, city officials can make better decisions about where to place adult businesses and how to minimize negative impacts on a community.

Next, a city could locate commercial and industrial zones where adult businesses might be accepted. A buffer area can also be established where adult businesses are not permitted—within a specified distance of commercial and industrial zones that are adjacent to sensitive land uses.

In addition to the land-use map, city officials can include the parcel map in their analysis, leading them to determine that no new adult business can be developed within five hundred feet of a parcel with an existing adult business.

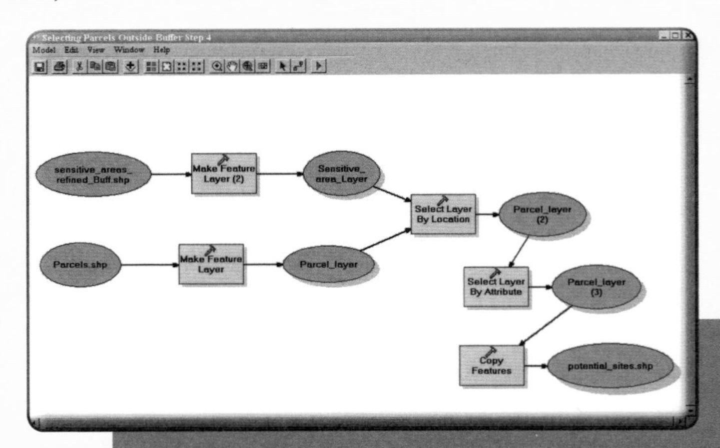

Figure 2.23 Chart shows the workflow process for selecting parcels outside a sensitive buffer area where adult businesses might be accepted.

Based on these restrictions or eliminations, cities can use standard GIS geoprocessing tools to create a map of possible locations where adult-business use will have the least negative impact on the community. This process is easily modeled in the GIS methodology. The model can be modified and run again for each change in the ordinance or in the land-use and parcel maps.

Facilitating public participation in decision making

3

Governments have encouraged the practice of public participation as a means of introducing citizen input before finalizing a decision or establishing policies. Initially, participation involved formal public hearings or public-comment periods where citizens could express their views and concerns on proposals brought to governing bodies. While this process was well intended, feedback was limited to those who attended and gained an understanding of the issues at hand. This resulted in some mistrust of government and claims of creating a new level of bureaucracy.

Governments have sought to increase public participation as a way to build citizen-government partnerships with the promise that the public's contributions would influence decisions. Progress has been made in encouraging and incorporating citizen feedback, with most of the gains realized by changing the ways this input was received. As a way to gain broader participation, governments have conducted town hall meetings; sponsored citizen committees, focus groups, and neighborhood meetings; published newsletters; administered mail campaigns; and promoted the use of the Internet. These forums have yielded a more informed public and lessened the sense that decisions were being made in a vacuum.

The integration of GIS technology provides for a true sense of collaboration. Most early uses of GIS software in the public-hearing process involved accurate notification of those affected by planning decisions. GIS technology also helped in the production of strong exhibits and provided a way to analyze and organize meetings and make adjustments according to the content, frequency, or location. These decisions were based on analysis that included information such as language spoken, educational background, or the number of individuals affected by a proposed policy change.

As the understanding of how GIS software could support public participation expanded, so did the applications of the technology in the public forum. The software evolved into a tool that could provide for immediate feedback to the citizens who chose to participate. GIS technology provided a means by which data could be instantly retrieved so the information could be brought to light. Often, the simple process of quick data retrieval lessened the number of meetings required, because the new information could be shared immediately.

GIS technology became a tool for modeling and performing "what if" scenarios based on the issues raised and enabling immediate presentation of the information to constituents. The Internet provided an even more powerful way to include citizens in the decision-making process. Citizens could log onto a Web site, draw boundaries for redistricting, and receive immediate feedback on their input without ever having to attend a public meeting or join a committee. As the use of GIS software has grown and data has become more accessible, citizens can present their cases via GIS output or analysis and present cases that staff might not have considered.

Cleveland advances into the high-technology sector with GIS

CHAPTER: Facilitating public participation in decision making
ORGANIZATION: Cleveland Division of Water
LOCATION: Cleveland, Ohio
CONTACT: Xander Mavrides, PhD, GIS manager xander_mavrides@clevelandwater.com
PROJECT: GIS Resident Workforce Program
SOFTWARE: ArcGIS
ROI: Increased efficiency, accuracy, and productivity, enhanced communication and collaboration

By Xander Mavrides

Perched along the banks of the Cuyahoga River and Lake Erie, Cleveland, Ohio, has ties to the water around it that helped establish it as a major manufacturing center early in the twentieth century. Since then, city leaders have sought to diversify Cleveland's economy and nearly a century later have stepped up efforts to cultivate a technology sector. In that vein, the City of Cleveland began strengthening its own information technology (IT) system, and in 2003 began an implementation of a citywide GIS.

The city's Department of Public Utilities, Division of Water, moved forward to develop an enterprise GIS to support the city's major business processes, including water engineering and permitting. This phase involved building several large GIS datasets from multiple sources, obtaining six-inch color orthophotos, and ultimately building the Cleveland Enterprise GIS group (CEGIS) to support the city's GIS.

To encourage public participation and provide an educated workforce for its expanding technology sector, the city decided to hire and train city residents to perform the massive GIS data-conversion tasks. Cleveland's GIS Resident Workforce Program was established to provide between thirty-five and fifty city residents with an employment opportunity that included extensive training in basic office applications, ArcGIS, data-conversion techniques, and data modeling. Ultimately, the participants in the program helped to convert more than 5,400 miles of water main, 450,000 service connections, 275,000 sewer laterals, 270 electric feeders, 600,000 parcels, and administrative boundaries into the GIS database

Figure 3.1 Cleveland's identity, and economy, is moored to the water around it.

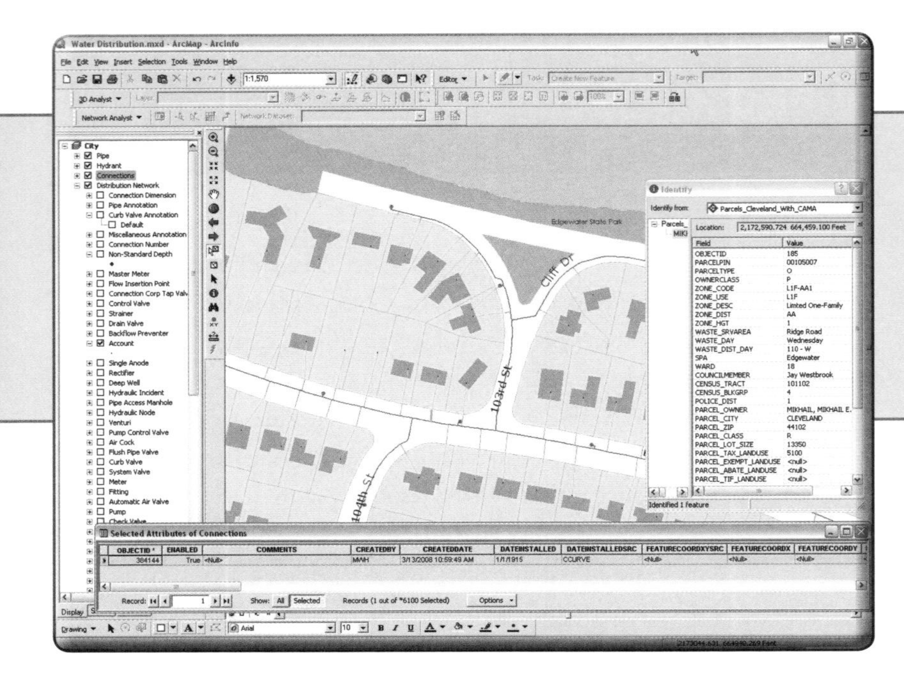

Figure 3.2 Cleveland's Department of Public Utilities, Division of Water, developed a citywide GIS, including this water-distribution network built in ArcMap.

Not only was the resident workforce staff reflective of the city's diversity, but citizens now also have an opportunity to further their careers by pursuing GIS and other related positions. In fact, CEGIS hired four program graduates as full-time GIS technicians.

In terms of data-conversion dollars, the costs of the conversion efforts were on par with other "off-shore" data-conversion companies. The true return on investment comes with the employment and training of Cleveland's residents and their contribution back to the local economy.

Figure 3.3 Cleveland Enterprise GIS Manager Xander Mavrides, left, stands with GIS Resident Workforce members who are now full-time employees with the city's enterprise GIS team. Next to Mavrides are, from left, Kevin Chubb, GIS technician; Lillian Coleman, permits and GIS technician; Justin Schaffer, GIS technician; and Anthony Atkins, GIS technician.

In 2007-08, Cleveland deployed several GIS applications, including hydraulic modeling, crime analysis, waste collection and snow-plow routing, a capital-improvement project-management tool, geoaccounting tools, and a webGIS for mapping, analysis, and reporting via enterprise databases. This would not have been possible without help from the Resident Workforce Program and CEGIS staff.

Not only was the resident work force staff reflective of the city's diversity, but citizens now also have an opportunity to further their careers by pursuing GIS and other related positions.

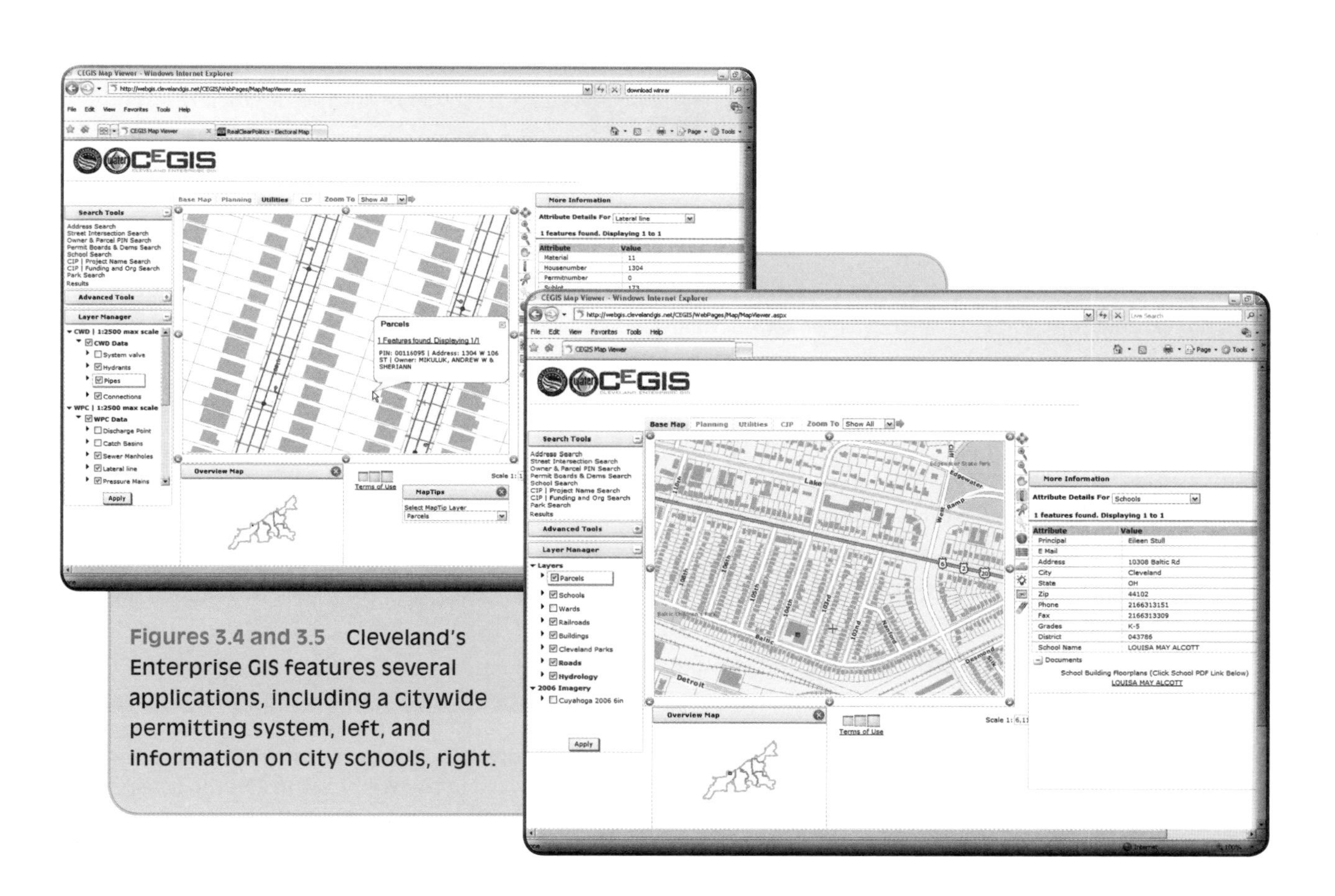

Figures 3.4 and 3.5 Cleveland's Enterprise GIS features several applications, including a citywide permitting system, left, and information on city schools, right.

Mapping derelict and sunken vessels along the coast of Georgia

CHAPTER: Facilitating public participation in decision making
ORGANIZATION: Georgia Department of Natural Resources
LOCATION: Brunswick, Georgia
CONTACT: Buck Bennett, Coastal Resources Division compliance and enforcement manager buck_bennett@dnr.state.ga.us
PROJECT: Locating sunken vessels
SOFTWARE: ArcIMS, ArcPad
ROI: Cost and time savings; increased efficiency, accuracy, and productivity

By Charles "Buck" Bennett

The Georgia Department of Natural Resources (DNR) has been struggling with abandoned and derelict vessels for several years. Based on recent estimates, it is believed that there are as many as one hundred sunken vessels along the Georgia coast. There have been sixty-five abandoned or derelict vessels identified as nonhistoric wrecks on the state's tidal-water bottoms. Nonhistoric wrecks are those vessels that have no significant historical value. These vessels include shrimp boats, abandoned recreational vessels, barges, and cranes. DNR's Coastal Resources Division (CRD) has taken on the task of locating, documenting, cataloging, and photographing these nonhistoric wrecks. The division is leading the public information effort in Georgia to identify wrecks and derelict vessels. This also includes the development of an ArcIMS Web site with a GIS interface to serve as a conduit between survey data from the field and stakeholders interested in the survey results. The Web site will be regularly updated to keep the boating public informed of these coastal hazards and to minimize the risk of collision and injury.

In many cases, there is no insurance coverage nor does the owner have the financial ability to retrieve the vessel. Usually, the owners of the sunken vessels cannot be found or ownership cannot be proven. Most boat owners escape financial responsibility either through bankruptcy laws or a clause in federal maritime law that limits an owner's liability to the value of the ship and its contents. Since most abandoned vessels are worthless, the owner's liability is usually zero. These abandoned and

Figure 3.6 The wreck of the Treasure D is located in the Wilmington River in Savannah, Georgia.

sunken vessels are made more hazardous because of the winding rivers, creeks, and tributaries and the high tidal amplitude along the coast of Georgia. Wrecks can be found in the marshes, submerged in tributaries near or in the Intercoastal Waterway, near bridges along Interstate 95, or in the sounds and coves.

Georgia's coastal marshlands encompass nearly 400,000 acres in a four- to six-mile band behind the barrier islands. These marshlands represent a considerable portion of the remaining marshlands along the entire eastern coast of the United States. Georgia's coastal tides average 7.2 feet in the marshes but exceed 10 feet during high spring tides (occurring during each lunar cycle). These high tides increase the dangers of water hazards such as sandbars, submerged wrecks, and structures. The 118 miles of Georgia's linear coast includes 2,400 miles of tidal tributaries and rivers and 184,000 acres of tidal waters.

Figure 3.7 Typical landscape shows Georgia's coastal marshlands.

Mapping sunken and derelict vessels with GIS

During the 2006 legislative session, monies were appropriated by the Georgia Legislature to remove some of the derelict vessels. A team was formed that includes members of DNR's CRD, Environmental Protection Division, and the Wildlife Resources Division's Law Enforcement Section to catalog, evaluate, and prioritize the vessels for removal. To facilitate and more accurately document the location of these nonhistoric derelict vessels, Buck Bennett, compliance and enforcement program manager, and Jill Andrews, acting operations program manager, applied for a grant through ESRI.

The Web site will be regularly updated to keep the boating public informed of these coastal hazards and to minimize the risk of collision and injury.

After receiving the Trimble Pocket PC and software from ESRI, staff members conducted a survey of Turners Creek in Chatham County near Savannah, Georgia. Turners Creek has become extremely congested with recreational sailboats and fishing vessels, along with derelict vessels. Along a quarter-mile section of Turners Creek, there are two known wrecks, a derelict fishing vessel, a public boat ramp with dock, and no less than seven recreational sailboats anchored, as well as a large marina, charter fishing docks, and dry-dock facility. This area of Turners Creek was the ideal location to begin collecting information and test the equipment.

Mobile mapping in the field

Staff traveled to the project area in Savannah, Georgia, in November 2006. In preparation for on-site mapping, the team loaded ArcPad into the Pocket PC and customized ArcPad with a pull-down menu containing the various information cells that would need to be collected in the field. The mobile GIS unit was then either carried into the field and placed on the deck of the small research vessel or hand-carried into the marsh and/or tributary to accurately mark the location of the sunken or derelict vessel. Information was entered through the drop-down menus on the Pocket PC.

Because of the large TIFF files and limited memory in the Pocket PC, staff members uploaded only those sections of coastal county maps necessary to conduct the survey of the project waterway. After collecting the data, staff members would download the updated point shapefiles in the office. A two-gigabyte secure digital (SD) card was added to the Pocket PC to enable the use of additional TIFF files. Photographs or side-scan sonar images were made of various wrecks to be used on the Web site to allow boaters to see the water hazard as it

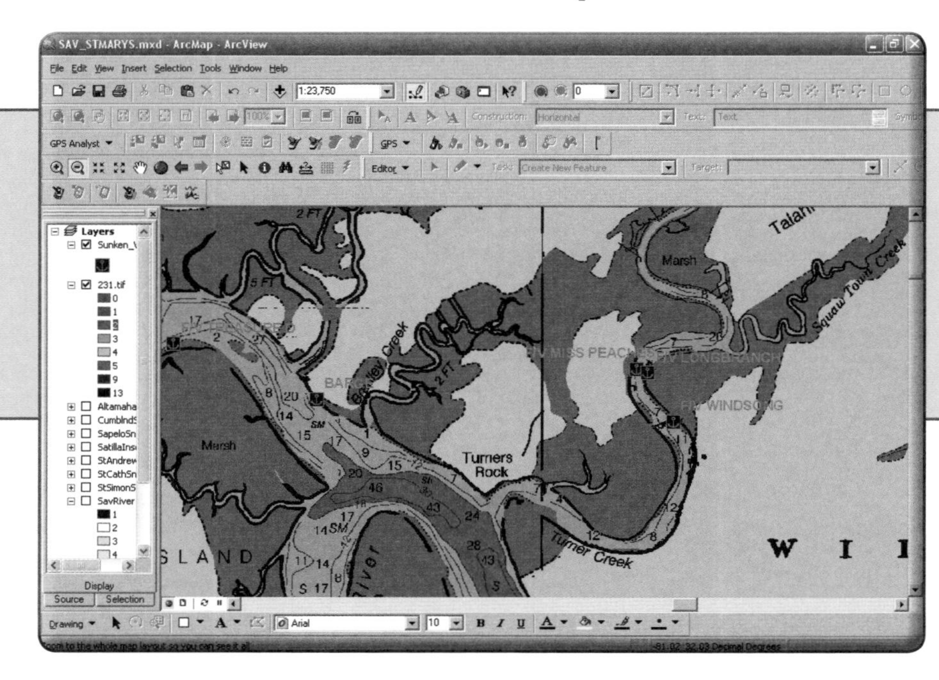

Figure 3.8 Screen capture shows Turners Creek and the Wilmington River in Chatham County, Georgia.

Figure 3.9 Side-scan sonar image shows a sunken barge in the Ogeechee River in Bryan County, Georgia.

exists. Because some of the derelict vessels have sunk in deep water and are not exposed, side-scan images are being used to enhance wreck site awareness.

These digital photos will be accessible via a link from an identified site on the finished ArcIMS Web site. Additional information, such as the impact to marsh vegetation from the vessel itself or from fuel and oil leaks, will be collected. This information will be analyzed using GIS software and made available to staff making decisions regarding the removal of these derelict nonhistoric wrecks. The site also will include a description of the marine debris, its location, and its condition.

Through newspaper articles, departmental publications, and personal contacts, the state of Georgia has recruited local boaters and charter fishermen to help identify additional wrecks or derelict vessels. The project will also assist staff, local governments, and possibly federal agencies in assessing marine debris as it relates to navigation, fishing, and environmental impacts.

The sunken-vessel removal process is based on public interest, navigational impact, relationship to shellfish and/or fishing, and economic impact. The Trimble equipment and GIS software enable more accurate documentation of the debris locations and their proximity to channels, marinas, high-traffic boating areas, reefs, shellfish harvest areas, and popular fishing areas. More than 90 percent of these wrecks are wooden and have a tendency to break up with age, weather, and wave action. The debris from derelict and sunken vessels becomes a floating or semisubmerged debris field, posing a hazard to boaters in Georgia waters. The U.S. Army Corps of Engineers and the state of Georgia are discussing the removal of some of the sunken vessels.

> **The debris from derelict and sunken vessels becomes a floating or semisubmerged debris field, posing a hazard to boaters in Georgia waters.**

Project completion

When the project is completed, the public will be able to access the sunken-vessel data through the Georgia Department of Natural Resources Web site. The test page can be found at http://dev.gadnr.org/dev/imf/imf.jsp?site=sunk. Some of the features have not yet been activated, and the site is an example of what can be expected when the project is completed. Because of bandwidth limitations, the background map will be composed of either navigational charts or 1:24,000 maps. The site will be updated regularly to add or remove derelict and sunken vessels.

Saving time

The handheld Pocket PC allows staff members to quickly collect GPS coordinates, log the information, and move to the next subject location. The CRD plans to continue to use mobile GIS in its fieldwork along with georeferenced photographs. Both technologies are an "information multiplier" for the public and researchers. This handheld technology significantly reduced in-the-field mapping time and has created more accurate information than with manual field-mapping methods or simply by using georeferenced aerial photographs alone.

Attributes of Sunken_VesselsUTM

FID	Shape*	OBJECTID	ID	NUMBER_	OTHER	VESSEL	COUNTY
0	Point	1	3	7	3836	BARGE	CHATHAM
1	Point	2	4	12	31503	F/V TREASURE D	CHATHAM
2	Point	0	0	0	0	fhncg	
3	Point	0	0	0	0	F/V WINDSONG	CHATHAM
4	Point	0	0	0	0	F/V LONG BRANCH	CHATHAM
5	Point	0	0	0	0	F/V MISS PEACHES	CHATHAM

Figure 3.10 Screen capture shows the attributes of the sunken vessel UTM.

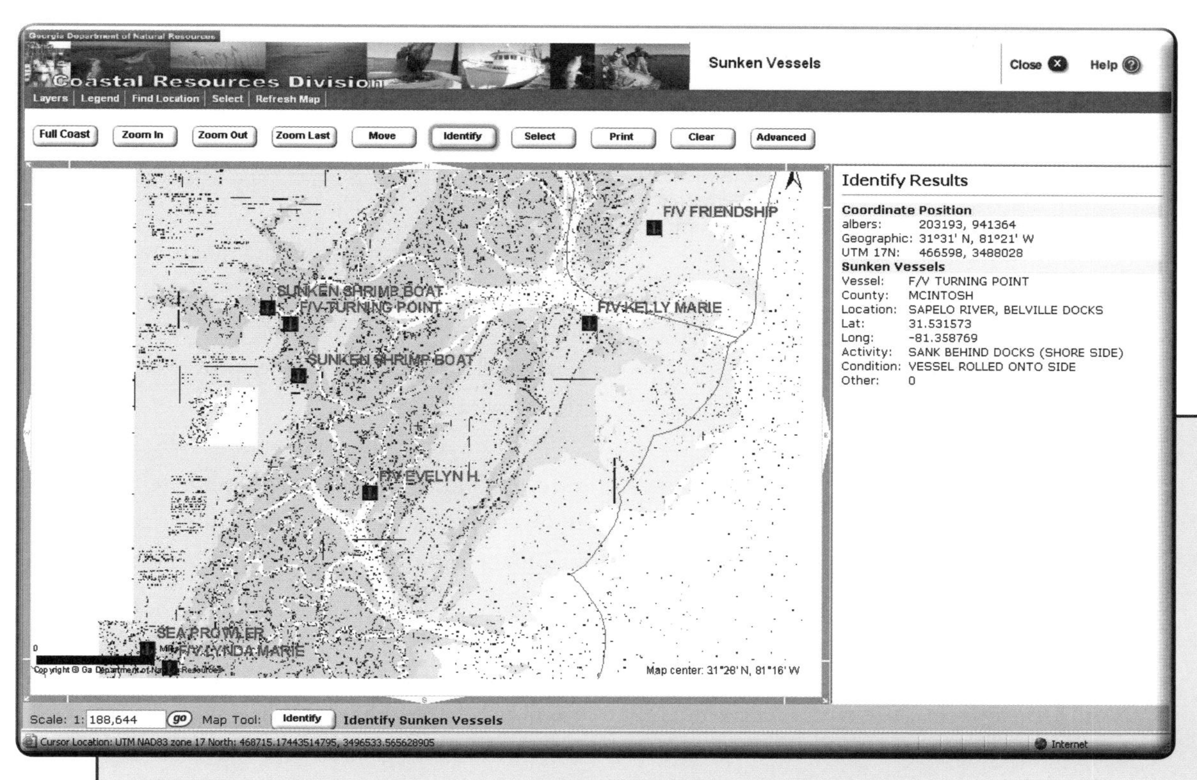

Figure 3.11 A Web page from Georgia DNR's Coastal Resources Division.

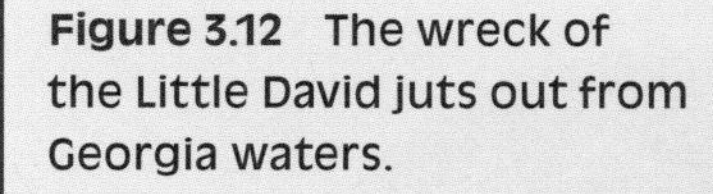

Figure 3.12 The wreck of the Little David juts out from Georgia waters.

Revitalizing Philadelphia's neighborhoods with GIS

CHAPTER: Facilitating public participation in decision making
ORGANIZATION: Philadelphia GIS Services Group
LOCATION: Philadelphia, Pennsylvania
CONTACT: Megan Heckert, business development manager, Avencia Inc. mheckert@avencia.com
PROJECT: Neighborhood Transformation Initiative
SOFTWARE: ArcSDE, ArcIMS, ArcGIS Server
ROI: Cost and time savings; increased efficiency, accuracy, and productivity; more efficient allocation of resources; improved access to information

By Bradley Breuer

State and local governments across the nation have invested substantial resources in developing enterprise-wide geographic information systems (GIS). The hardware, software, database development, and training required for these systems to mature have required major commitments from government, both in terms of finance and personnel. In many places, the progress has been quite remarkable. From streets and sewers to epidemiological and real estate information, our state and local governments are using GIS-centric workflows to catalog and manage myriad data sources. Migration to versioned geodatabases, the resolution of disorganized spatial data, and enhanced data integration have all been the hallmark of the most successful state and local governments, leaving many to anticipate what lies ahead for GIS. How do we prove the return on these investments?

With a backbone of strong GIS infrastructure, including the necessary software, hardware, industry knowledge, and data integrity, there is abundant potential for using these technologies to enhance the way that state and local governments provide services. Governments can leverage spatial-data warehouses to make decisions in every domain from public health, property management, and emergency management to citizen services, crime and fire, and a range of other areas. New Web-enabled GIS technologies and advancements in the functionality of these technologies have opened the door to putting mapping and spatial analytics at the center of every public-policy discussion and in the hands of every decision maker.

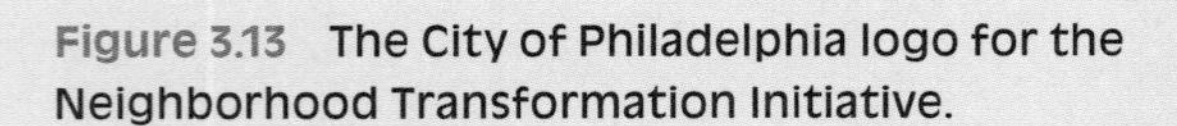

Figure 3.13 The City of Philadelphia logo for the Neighborhood Transformation Initiative.

New Web-enabled GIS technologies and advancements in the functionality of these technologies have opened the door to putting mapping and spatial analytics at the center of every public-policy discussion and in the hands of every decision maker.

In Philadelphia, the GIS Services Group, part of the mayor's Office of Information Services, has used its GIS for decision support across various departments and business processes. Specifically, two Web-enabled GIS applications have enhanced the community and economic development efforts of the mayor's Neighborhood Transformation Initiative (NTI). The city's suite of geospatial Web services and enabling technologies are products of the next generation of thinkers and decision makers who want to make more of their GIS investments. Moving GIS to the next level requires a renewed sense of purpose and a willingness to think creatively about how state and local government can be more effective using Web-enabled GIS applications for decision support.

The GIS backbone

The City of Philadelphia has made a considerable investment in the design and development of a Unified Land Records System (ULRS). The initial phase of development focused on the creation of a seamless map layer of real estate parcels. The most recent phase has focused on the development of an

Figure 3.14 The City of Philadelphia has used GIS to advantage to further economic development.

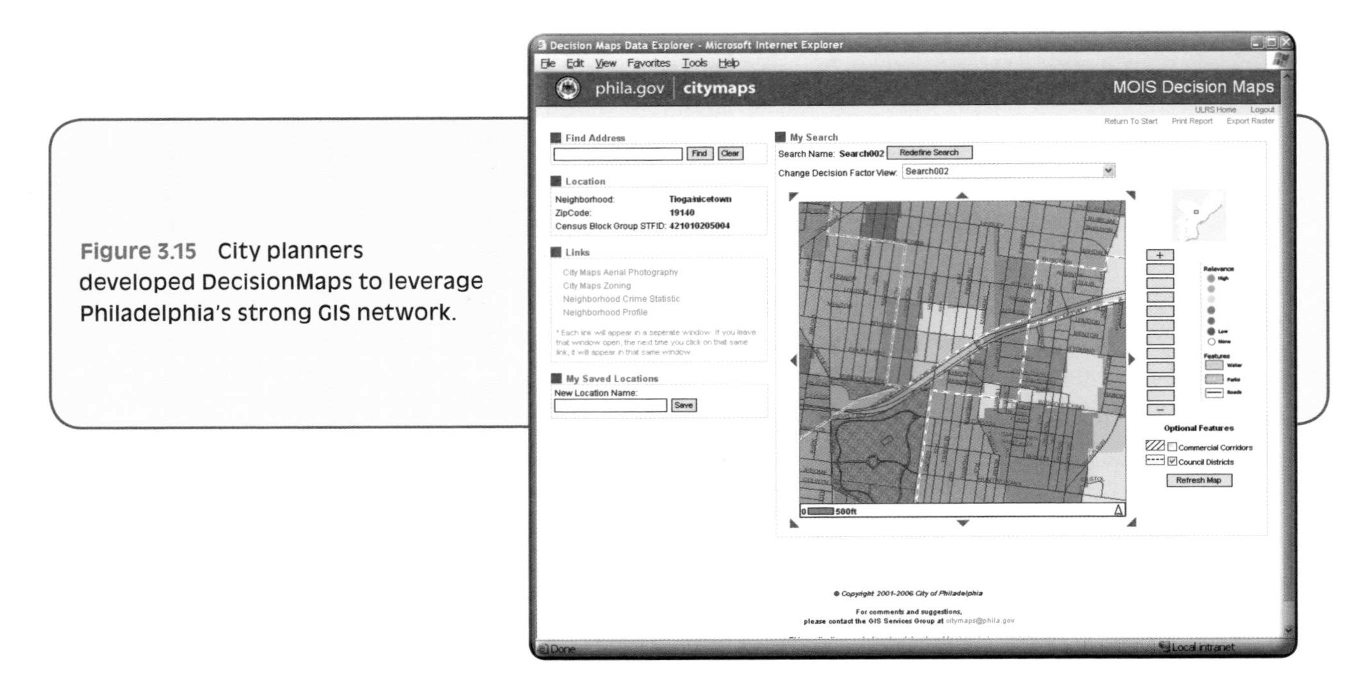

Figure 3.15 City planners developed DecisionMaps to leverage Philadelphia's strong GIS network.

address-integration system that overlays the parcel layer with other important geographic data such as inspections, revenue records, land title records, and tax accounts. The city also maintains hundreds of other spatial and nonspatial datasets related to the business practices of various city departments. Many of these efforts are coordinated by the GIS Services Group, with ArcSDE, ArcIMS, ArcGIS Server, and Oracle running the city's spatial-data services and applications.

In April 2001, the City of Brotherly Love launched the NTI, a multifaceted, $300 million effort to reverse a decades-long history of blight and disinvestment. NTI outlined a provocative strategic action plan to tackle blight and stimulate investment in Philadelphia neighborhoods. Information technology, specifically GIS, has played a critical role in developing policy for the initiative, making NTI a significant user of the city's GIS. According to Eva Gladstein, director of NTI, "GIS has played an important role in Mayor (John F.) Street's NTI. Decision tools have helped make, analyze, and interpret policy and guide resources and investments. The city has embraced these methods and will be expanding the role and use of these tools to provide more effective service to the citizens of Philadelphia."

DecisionMaps

Where is the ideal place to locate a business or target community and economic development efforts? Where is the most appropriate site for a new high school to serve a specific region or to place a community service and recreation facility? Should light industry locate in the former Philadelphia naval yard in the south of the city or in North Philadelphia close to major transit lines and former brownfield sites? Where is the best street corner for a

Figure 3.16 DecisionMaps uses GIS technology to help businesses pick the right location.

specialty flower shop to locate based on population demographics and proximity to business corridors? Where is the best place to locate a business that targets young people, a concentration of moderate- to high-income households, and renter-occupied housing? Is that place also located near public transit lines? The answers to these questions are critical to making business-siting decisions and decisions about where to prioritize government efforts in community and economic development.

To help with the answers, Philadelphia planners developed DecisionMaps, a Web-based business-siting tool. Location has an enormous impact on the success or failure of a business, underlining the importance of making good choices. These choices are ultimately based on geography, proximity, density, and demographics. Our intuitive sense of where to site a business accounts for various geographic factors, with some factors having more influence than others. DecisionMaps takes a user-centered approach, couples it with a map interface, and lets the user iteratively create site suitability maps.

"In developing DecisionMaps," said Jim Querry, GIS director for the city, "we sought to leverage the city's strong GIS infrastructure and give users advanced tools for economic development planning and a tool for outreach." Users can access DecisionMaps in a Web-based browser with a user-friendly interface. They are given a menu of geographic priorities, such as proximity to transit lines, proximity to business districts, and whether

Non-GIS personnel are able to use DecisionMaps to prioritize funding for commercial corridors across the city based on need, existing resources, and performance of past funding initiatives.

the site is in an economic incentive zone or wireless Internet coverage area. There is also a menu of various demographic factors, including age, per capita income, and educational attainment. Based on this menu of decision factors, users weight their preferences by giving some factors high "prefer" scores, other factors low "avoid" scores, and leaving other factors neutral.

After weighting these concerns, users are presented with a map showing hot spots or the most desirable business-siting locations. They can pan or zoom in on the map from a citywide overview to a street-by-street analysis. Each 100-x-100-foot cell has a unique score based on user-designated preferences. DecisionMaps uses ArcGIS Server and ArcGIS Spatial Analyst to calculate multiple decision scenarios. Using the ArcObjects framework in an iterative manner, the user can quickly return to the preference menu, reweight inputs, and then see the new results in an interactive map. The system currently has more than fifty decision factors and is also designed to enable designated administrators to add new geographic layers to adapt to changing economic development needs and available data resources.

The NTI uses DecisionMaps to plan allocation of funding for a proposed bond issue to invest in the improvement of neighborhood commercial corridors. Non-GIS personnel are able to use DecisionMaps to prioritize funding for commercial corridors across the city based on need, existing resources, and performance of past funding initiatives.

Putting BUILD to work

What is the next step after users find the right neighborhood, street corner, or section of the city to locate? Where do they find reasonably priced land, and how do they aggregate several parcels into a single piece of land that is developable for their business needs? How does the city put undeveloped land in the hands of private developers and community development corporations? Providing a better link between the city's repository of spatial and administrative data on property and those who want to acquire property for development would provide more efficient economic development in Philadelphia.

Building Uniformity in Land Development (BUILD) is an application focused on serving this practical need. Using a Web-based search interface, users can search for properties and view cross-agency details in both textual and spatial formats. BUILD users can search for single or multiple properties by address or conduct an area-based search for properties that meet certain development criteria. They can scroll through textual search results or view an interactive map of the property search result.

Central to the BUILD application is a property Operational Data Store (ODS), which serves as a near real time central database repository of property-related administrative data collected from a

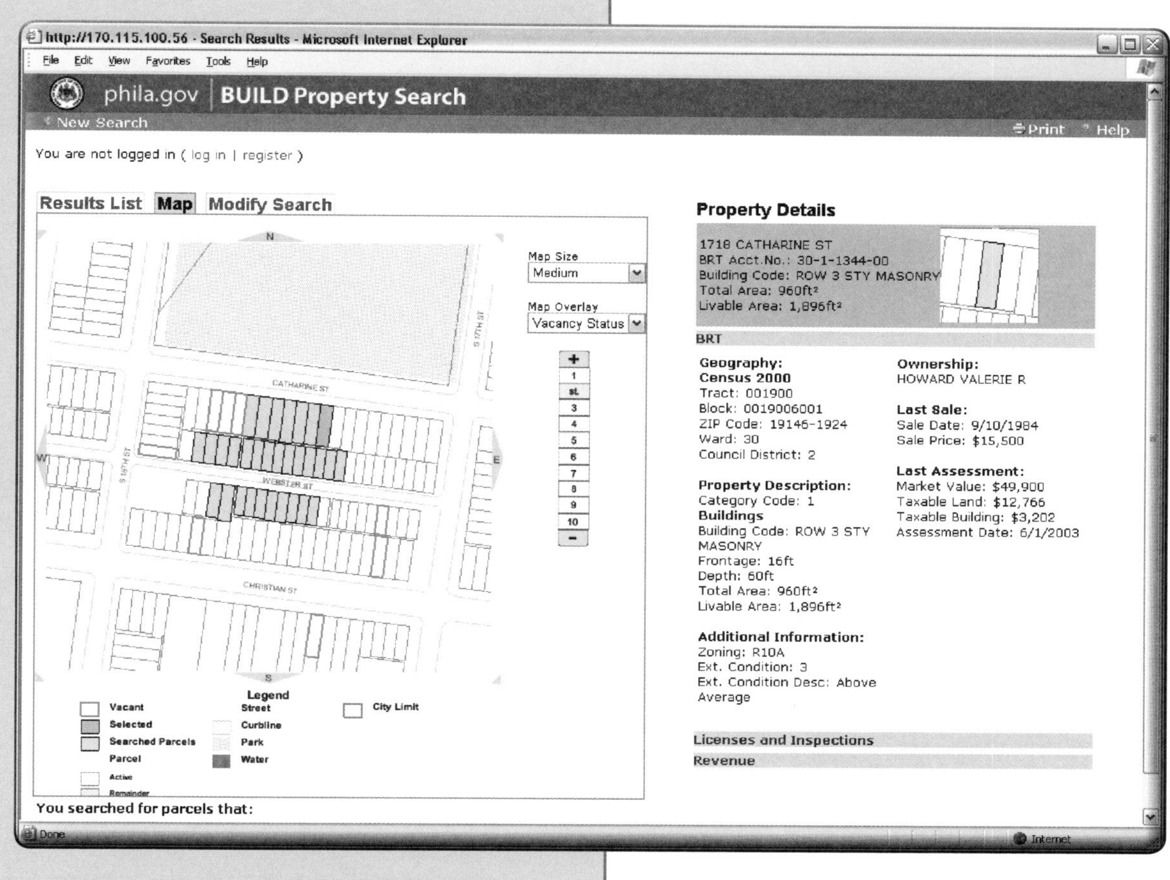

Figure 3.17 Philadelphia's BUILD application provides developers with useful information for transforming city neighborhoods.

variety of city departments and agencies, including the Board of Revision of Taxes (BRT), Department of Records (DOR), Department of Licenses and Inspections (L&I), Department of Revenue, and the Water Revenue Bureau (WRB). ODS also provides an inventory of publicly owned properties from various city agencies, enabling policy makers to intelligently analyze the public inventory and begin marketing and disposing of surplus inventory to interested developers. Phase two of the BUILD application will permit users to select properties online and submit a request for the acquisition or disposition of those properties to the appropriate agency.

BUILD and DecisionMaps are complementary counterparts in the city's overall strategy to engage businesses in revitalizing Philadelphia neighborhoods. By providing advanced analytical tools that support location decisions and supplying a data-enriched mapping environment for property exploration, the applications work in harmony to leverage GIS resources previously unavailable to businesses in a user-friendly manner.

GIS and dynamic Web services

A Web service is a technology that enables the publishing of dynamic information via the Web. These applications are generally prebuilt and ready for a department to simply plug into its existing Web site, application, or database. Within an organization, Web services can serve hundreds of users and be implemented rapidly in a straightforward way. The City of Philadelphia describes its set of geospatial Web services as a toolbox of reusable GIS components. The city has developed common GIS functionality such as geocoding and map generation as a set of tools that can be deployed over a network to any department. Hosted by the GIS Services Group, the city currently offers the following Web services:

- Geocoding and batch geocoding
- Reverse geocoding from a GPS point
- Street address standardization and geocoding to clean poorly formatted addresses and locate the corresponding geographic coordinate
- Map generation
- Create an image file or PDF to display any location in Philadelphia
- Look up service area for a range of services like rubbish collection days
- Determine the ZIP Code, police, fire, or council district for any street address and find nearest facilities
- Locate the libraries, recreation centers, health centers, or schools nearest any address
- Find all known addresses associated with a parcel
- Retrieve a list of all addresses from other department systems related to a given parcel
- Look up land records by address or account number
- Use any known address or account number to retrieve land record data from any address-integrated database

This suite of geospatial Web services played a critical role in developing the applications. Without this infrastructure in place, the pace of development and richness of features in the applications would have been seriously hindered.

Did it pay off?

DecisionMaps helps city departments make decisions about where to target government initiatives and interventions, attract new business, and support the growth of existing Philadelphia businesses through the NTI. The DecisionMaps methodology has directly supported the agency's demolition and site-selection efforts. These applications have exposed GIS to a range of users who normally would not use mapping and spatial analytics as part of their policy- and decision-making practice. Beau Bradley, a senior GIS analyst for NTI and a user of the DecisionMaps application, says, "We sought to help nontechnical decision makers to better process information in their planning and policy efforts by hiding the technical programming behind the application and focusing on the geographic outputs."

Geospatial Web services have saved the City of Philadelphia hundreds of hours of GIS developer time. Senior GIS developer Clinton Johnson has used Web services to add mapping and spatial analytics to dozens of Web sites and workflows within the city's infrastructures. "Web services have enabled the GIS Services Group to deploy geospatial functionality more quickly and more effectively," Johnson said. "It has eliminated redundancy and enabled the city to use prebuilt GIS functionality over and over again, each time impressing a new city department with fast and effective GIS integration." An example of this fast integration was the introduction of mapping to a new Fairmount Park Web site in a matter of days, enabling residents and visitors of Philadelphia to view data-rich maps of America's largest urban park.

It has eliminated redundancy and enabled the city to use prebuilt GIS functionality over and over again, each time impressing a new city department with fast and effective GIS integration.—Clinton Johnson, senior GIS developer

Acknowledgments

This article was a collaborative effort between the City of Philadelphia mayor's Office of Information Services (GIS services group), the mayor's NTI, and a special project team working on the BUILD application for the city.

Purchasing development rights in Washington County, Wisconsin

CHAPTER: Facilitating public participation in decision making
ORGANIZATION: Washington County Planning and Parks Department
LOCATION: West Bend, Wisconsin
CONTACT: Eric Damkot, GIS manager eric.damkot@co.washington.wi.us
PROJECT: Land preservation
SOFTWARE: ArcEditor, ArcIMS, ArcInfo, ArcSDE ArcView
ROI: Increased efficiency, accuracy, and productivity; enhanced communication and collaboration

By Eric Damkot

Agriculture historically has been the basis for much of the local economy of Washington County, Wisconsin, which covers 435 square miles, has a population of 126,000, and is just northwest of Milwaukee. A 2004 study by the University of Wisconsin Extension shows that agriculture in Washington County supports nearly 5,000 jobs and annually contributes more than $629.9 million to the local economy.

The county has experienced rapid growth with a population increase of more than 97 percent from 1970 to 2005. This growth, combined with trends toward using larger land parcels for each household, lower numbers of people per household, and a common desire to live near open space, has created tremendous development pressures on the rural landscape in Washington County. From 1970 to 1995, land converted from rural to urban land uses averaged 1.2 square miles annually. From 1995 to 2000, the rural-to-urban conversion rate increased to 2.3 square miles annually.

In 2005, the Washington County Planning Conservation and Parks Committee recommended the formation of a purchase of development rights (PDR) task force in the interest of preserving the remaining large contiguous tracts of prime agricultural lands in Washington County. The thirteen-member task force consisted of local farmers, interested citizens, local officials, county supervisors, and representatives from local agencies, land trusts, and conservation organizations. Washington County Planning and Parks Department staff provided administrative and technical support. The task force's mission was to analyze the potential for a PDR program in Washington County and to develop a strategic plan of action to implement such a program.

Although PDR has been successfully used in the United States since 1970, and twenty-seven states currently have a PDR program, its use in Wisconsin has been limited. Currently, there is no state program, no county program, and only two municipal PDR programs in Wisconsin. Under a PDR program, a landowner voluntarily sells his or her rights to develop a parcel of land. The property remains privately owned, on the local property tax rolls, and the landowner retains all other ownership rights attached to the land. The landowner can

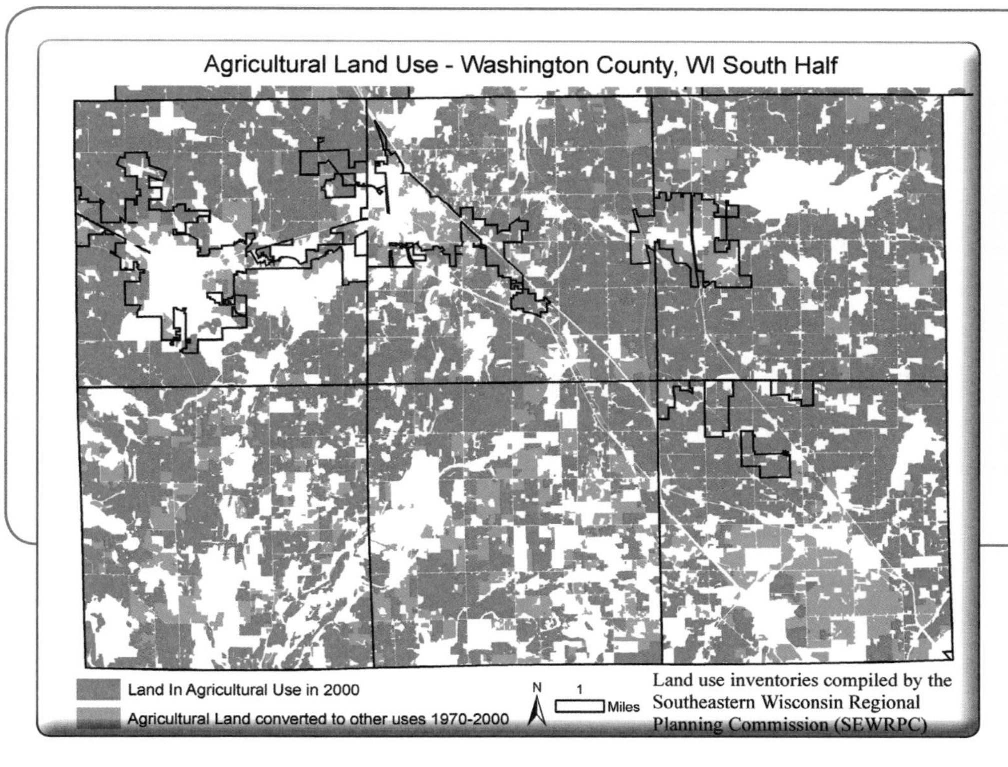

Figure 3.18 Land converted from rural to urban uses averaged 1.2 square miles annually from 1970 to 1995. From 1995 to 2000, the rural-to-urban conversion rate nearly doubled, to 2.3 square miles annually. Development pressure is greatest in the southern part of the county.

sell or transfer the land, but all development restrictions are transferred to the future owners. The PDR easement generally reduces the market value of the land, making it more affordable for local farmers to purchase. The price paid for the development rights easement is typically based on the number of dwellings permitted under current zoning or the difference between the current market value and the value of the land if used exclusively for agriculture.

Identifying prime agricultural land

When studying the feasibility of this kind of program, the PDR task force needed to identify priority areas. The priority areas had to be both functional and easily defined. Washington County's ongoing investment in high-quality GIS data, software, and staff combined to provide the thorough analysis the PDR task force needed to achieve this goal.

Washington County's investment in GIS began with a pilot project in 1994 and has grown into a three-person GIS division within the Planning and Parks Department. The GIS division is responsible for maintaining cadastral, roads/addressing, and shoreland zoning/hydrography datasets in addition to data for numerous smaller projects. Although contained within the Planning and Parks Department, the GIS division has completed projects for nearly every Washington County department and some local municipalities. In addition to the GIS division, numerous county departments and other divisions within the Planning and Parks Department have staff that use GIS as part of their job. Washington County has a mix of ArcView, ArcEditor, and ArcInfo licenses. ArcSDE and Microsoft SQL Server are used for data storage and ArcIMS

As the appointed task force worked to question the original assumptions, the GIS could efficiently analyze how those changes affected the result. Because all scenarios were fully investigated, the task force could more easily reach a consensus.

for broader data distribution. As the appointed task force worked to question the original assumptions, the GIS could efficiently analyze how those changes affected the result. Because all scenarios were fully investigated, the task force could more easily reach a consensus.

Prime agricultural areas can be defined in numerous ways, so the PDR task force felt it was important to draw from previous efforts so that the data used was reputable and defendable. In 2004, the Washington County Board of Supervisors approved an update to the Park and Open Space Plan for Washington County. Included in this adopted plan was a map showing the county's prime agricultural lands as mapped by the Southeastern Wisconsin Regional Planning Commission. This map served as the starting point for identifying priority areas for the potential PDR program.

In addition to the prime agricultural lands identified in the Park and Open Space Plan, two

Figure 3.19 Agriculture in Washington County supports nearly 5,000 jobs and is part of a diverse local economy.

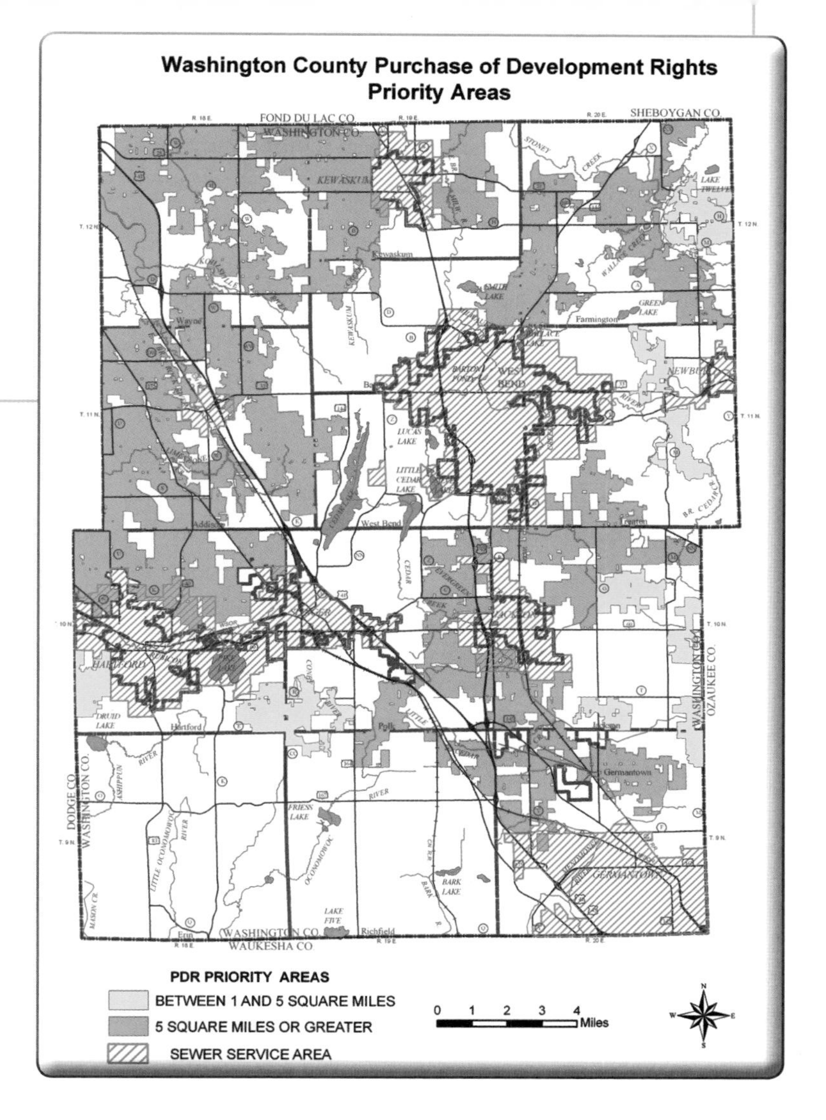

Figure 3.20 Preliminary priority areas for a PDR program in Washington County were identified through a GIS analysis that leveraged information from a variety of existing datasets.

townships also had updated their farmland preservation plan. A farmland preservation plan qualifies land within an exclusive agriculture-zoning district to qualify for Wisconsin's Farmland Tax Relief Credit program. For these two townships, the data from their farmland preservation plan was substituted for that of so-called prime agricultural lands.

Even though the data came from reliable sources, it was already a few years out of date. Recent annexations and developments were not properly reflected. To remedy this problem, the subdivisions, condominiums, and certified survey maps from the county's cadastral dataset were clipped from the prime agricultural base. The base was further modified to eliminate those areas recently annexed to any of the incorporated cities and villages. Completing this step eliminated substantial tracts in the more rapidly developing southern portion of the county that would have qualified for the PDR only a few years before. Seeing these areas removed from consideration helped clarify the need for a PDR program to preserve remaining agricultural land.

To get the most value from the county's potential investment in a PDR program, the PDR task force wanted to identify large continuous tracts of farmland. Thresholds of one square mile and five square miles were set. All prime farmland that was not part of these larger tracts was removed from consideration. During this analysis, it became apparent that some areas were removed from consideration because prime agricultural lands were isolated from one another by lands already in public holdings. To alleviate this problem, the cadastral dataset was queried to retrieve all parcels currently owned by the state, county, or local municipalities. The publicly owned areas were then added to the prime agricultural dataset. The prime agricultural lands were then requeried to extract the areas greater than the one-square-mile and five-square-mile thresholds to determine the priority areas included in the final PDR task force report.

Other data was also considered during the planning process—most notably, planned municipal sewer service areas. Tracts within the sewer service area are logical areas for presumed future growth in the county. It was debated whether the entire sewer service area should be excluded from the PDR priority areas. Ultimately, the task force elected to show sewer service areas as an overlay and unilaterally eliminated these sewer service areas from PDR priority areas.

The result from the analysis identified places in the county where large tracts of contiguous prime agricultural land still exist. These areas were included in the PDR task force's final report as preliminary priority areas. The analysis and corresponding maps helped citizens and elected officials understand the rate at which Washington County is losing its rural character and the preservation opportunities that are still available.

In March 2006, the Washington County Board of Supervisors passed a resolution creating the first countywide PDR program in Wisconsin. GIS will help determine applicant eligibility and prioritize the best use of limited funds during the program's implementation. As the program matures, GIS will track where development rights have already been purchased. A successful PDR program will help ensure that agriculture remains an important part of a diverse local economy. The open spaces and rural character that attract many people to Washington County will be preserved in perpetuity.

The analysis and corresponding maps helped citizens and elected officials understand the rate at which Washington County is losing its rural character and the preservation opportunities that are still available.

Exercise in decision support

GIS technology strengthens residents' case against adding a liquor store

A once peaceful, slow-growing community is experiencing rapid residential and retail development. Each week, developers submit their plans to the city and attend planning commission and city council meetings with the hope that their investment will move forward with minimal delays. Most of the developments are approved with no changes and no objections from residents and businesses.

In areas targeted for economic development and much-needed services, investors are looking to capitalize on the community's growth by developing infill projects that are primarily small industrial buildings and strip retail centers.

In this scenario, a small retail center was approved and developed in a neighborhood with large older homes. The area had been devoid of heavy traffic and retail offerings. While local residents assumed that the lot, vacant for more than fifty years, would eventually be developed, they were in no hurry to see changes to their neighborhood.

The small development company took a considerable amount of time to complete the project. Competition for prime businesses was tough, and most of the newer businesses were opting for locations closer to the new residential tracts. Filling the vacancies was slow—first, a restaurant, then, a furniture store. Neither venture caused alarm to homeowners who carefully watched as businesses were added.

But then along came a request to open a carryout liquor store. The notice of the application for a liquor license caused residents to mobilize to stall the addition of such an establishment. While the council and

Figure 3.21 The city council won rejection of yet another new liquor store in the community based on GIS analysis that showed there was already a plethora of liquor stores in that census tract.

staff had concern about the addition of a liquor store, the land use was permissible, and the state Department of Alcoholic Beverage Control dutifully proceeded with the application process for a liquor license. The main concerns of residents, meanwhile, were the threat of a "stop and rob" market and the attraction of undesirable clientele.

Under these conditions, elected officials and staff would be obliged to accept the business and liquor license based on city zoning and feedback from the state, which would inevitably result in upsetting residents who have come to love living in this quaint community. Elected officials and city staff would be left to hope that residents' fears were unsubstantiated, and that the business would become a welcome addition over time.

How can a city use GIS to respond to opposing viewpoints between residents and businesses? How can GIS help staff embrace public input and protect a community's identity?

The GIS difference

Public hearings provide residents with the opportunity to express their concerns about a policy or decision that is facing an elected body. Often, citizens at such public hearings present petitions aimed at swaying the vote coupled with emotional pleas to cast a vote in their favor. While these presentations are heartfelt, they do not necessarily offer valid justification for overruling existing policies and regulations.

However, in the case of this liquor store, one of the community leaders who rallied the residents had extensive experience using GIS. While petitions and emotional pleas were presented, this leader came armed with a fact-based GIS analysis and valid arguments to present to the council.

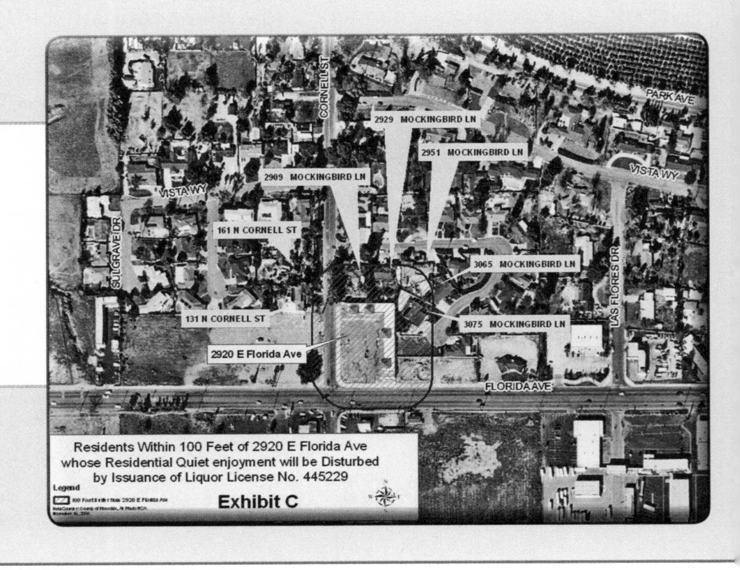

Figure 3.22 A 100-foot buffer shows residents who would be affected by the addition of a liquor store.

Through research, the community leader first found that the Department of Alcoholic Beverage Control approved liquor licenses based on the number of existing liquor licenses within a census tract. He then used GIS software to map all the carryout liquor licenses within his census tract and surrounding census tracts. In addition, a multiring buffer analysis was done from the proposed location of the liquor store.

With the GIS software, he was able to show the relationship between homes, churches, schools, and the proposed liquor store. What he was able to visually communicate to the council was that, contrary to what the Department of Alcoholic Beverage Control stated, there was already an overabundance of liquor stores within the census tract, and the approval of the business did not meet the ruling of a needed service.

The council recognized that the facts presented via GIS technology provided a solid foundation to deny the business. The result was a letter from the council with the accompanying documentation to the Department of Alcoholic Beverage Control requesting denial of the liquor license based on the findings. More important, the council found value in public input. GIS helped the council protect a neighborhood and better serve citizens.

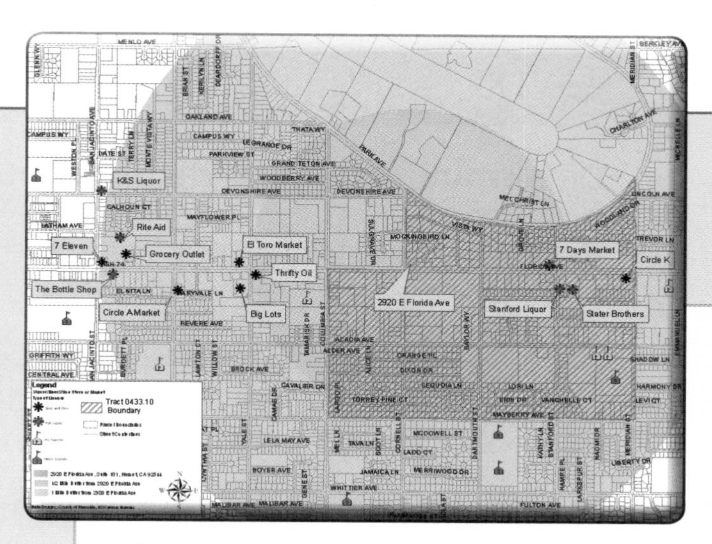

Figure 3.23 A multiring buffer drawn around a proposed new liquor store shows all of the carryout liquor licenses already located close by.

Figure 3.24 A presentation based on GIS technology convinces the city council to deny an unwanted business based on the facts rather than just emotions.

Exercise in decision support

Determining special-district boundaries

The formation of special districts or other kinds of service districts requires local governments to collaborate with many types of stakeholders, both public and private. Determining the boundaries of a proposed district can be a politically sensitive process.

Special districts often overlap or are adjacent to other political and jurisdictional areas where administrators have a vested interest. For instance, a school district might extend across two or more townships with dissimilar demographic profiles. In some cases, certain jurisdictions might not permit an overlapping function. For instance, a county's mosquito abatement district might not be allowed to impinge on federal lands.

Special districts are usually created through legislation or a public referendum. When this happens, the limited government services granted to the district may have an unexpected impact on local service-providing departments, such as police and fire, and sometimes more than one municipality is affected. Additionally, taxes and fee programs established within the district may unfairly burden low-income groups.

The successful implementation of a special district requires input from all stakeholders. Government organizations need to have the ability to share plans with citizens and incorporate feedback from the public and others into those plans. How can local government effectively create an ongoing dialog about a proposed boundary with public and private stakeholders?

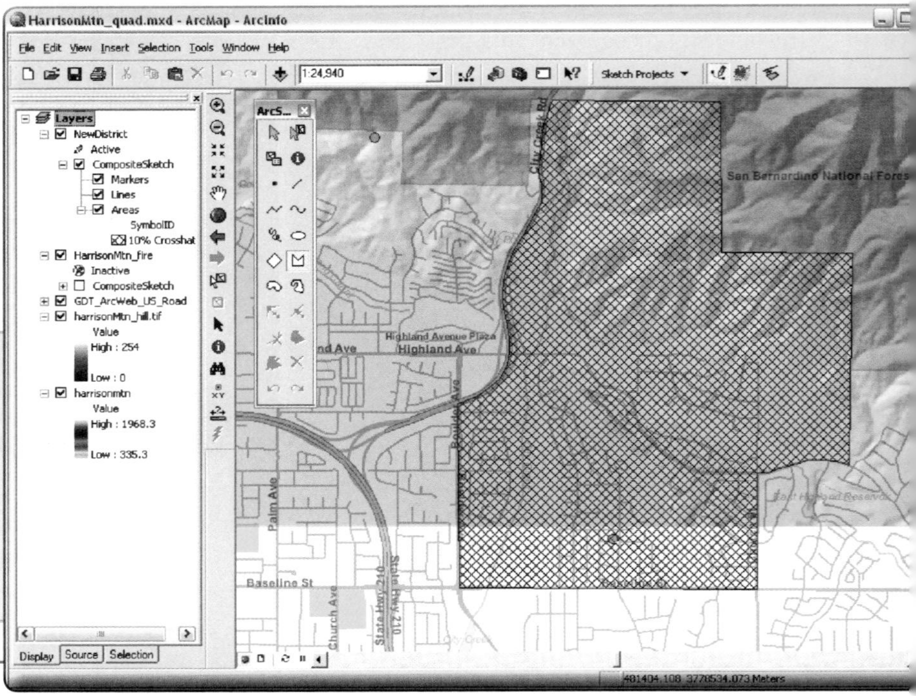

Figure 3.25 A proposed district boundary is drawn using ArcSketch. The district is shaded with crosshatch marks.

The GIS difference

From the outset, local governments can use GIS tools to improve special-district planning. A sketch of the proposed district boundary can be drawn directly on a map displayed on a computer screen with ArcGIS Desktop software. Sketching is a fast way to express an idea in GIS methodology and doesn't require GIS expertise—the software handles data-management functionality.

The sketching environment can be set to allow snapping to other GIS layers. Snapping, an automatic editing operation that connects coordinates of points or features within a certain distance of one another, can help shape the sketch so it follows the edges of other boundaries or street centerlines. A manager can also draw freely, guided by the terrain features shown in background layers such as hillshade relief or aerial imagery.

Once the sketch of the proposed district is complete, it can be saved and incorporated into other maps and analyses. Since a GIS sketch is real data, it can be used to find out what other jurisdictional boundaries intersect or touch the proposed boundary, the area of the district, or the number of miles of city streets contained within the boundary. The proposed boundary can also be overlaid with census data to create a demographic profile.

The sketch can be published to a GIS server, which makes the sketch available as a map service to Web mapping applications. That means the information can be added to an online map or globe and visualized within a reliable context.

For example, stakeholders can use their Internet connection to add the sketch service to an application like ArcGIS Explorer, where they can zoom in on the proposed district, rotate around it, and tilt the landscape to investigate it from a 3D perspective.

Presenting proposed boundary information on a map allows stakeholders to review the planned special-purpose district within a geographic context and gives officials an opportunity to investigate potential political entanglements.

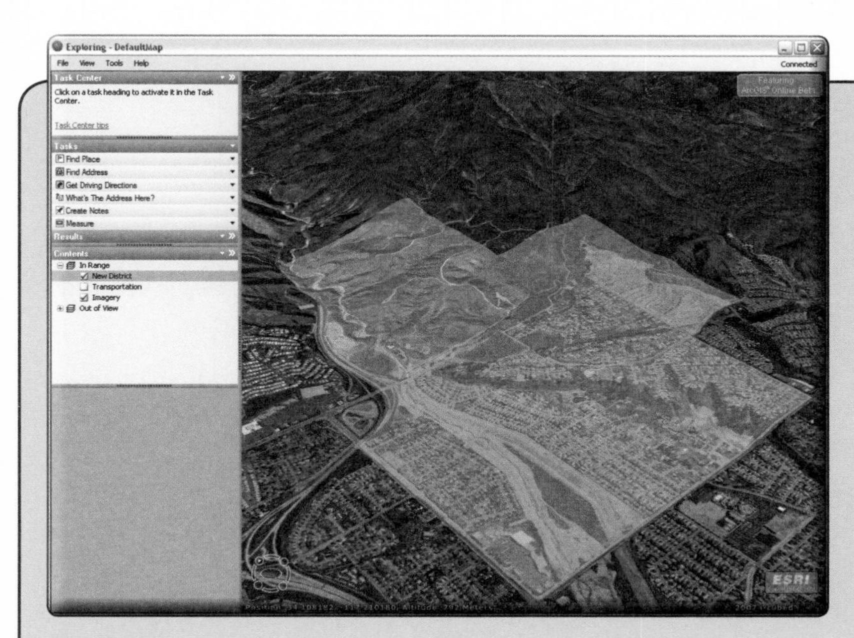

Figure 3.26 After being published to a GIS server, a sketch service can be added to a Web mapping application such as ArcGIS Explorer.

Making decisions under pressure

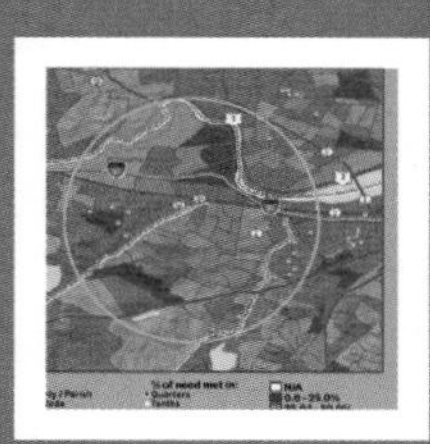

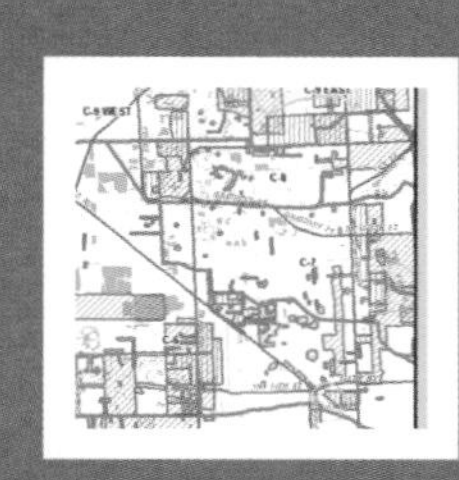

4

For most public-policy experts, having to make decisions under pressure comes with the job. Because the decisions public officials make unquestionably impact public finances and disparate groups with varying expectations, public scrutiny almost always comes into play as each decision is made.

Making decisions under pressure is not always associated with an emergency situation where time is critical and lives could be in danger. In fact, public officials are constantly faced with making decisions under pressure when emergencies are not imminent. From deciding to close or build a public facility such as a fire station to studying and developing fees supporting public services to establishing policies and ordinances that will lessen the impacts of global warming, the pressure to make a decision is usually present in some respect.

Decision making under stress means working in an environment of uncertainty characterized by shifting or competing goals, time constraints, high stakes, multiple players, or ill-defined problems. To make matters worse, these types of decisions are often made with limited public input or are forced by a vocal minority who might want to influence a decision. Public leaders must be ready to move away from groupthink where the voice of a few or the persuasiveness of a single thought takes rein without all of the options being weighed.

A GIS that supports public policy can serve to eliminate reactive decision making by involving all interested parties in the process and building a compromise. Most importantly, the GIS process seeks to reduce emotional responses through fact-based presentations.

While GIS is a highly visual tool, it is more importantly a statistical and modeling tool. GIS technology supports public policy with information based on science, professional perspective, modeling, scientific measurements, and predictive prototypes. For example, GIS software can integrate demographics, demographic projections, psychographic groupings, and opinion polls to reach a consensus.

As emotions diminish and the pressure for immediate action is minimized, the focus can center on reasoning things out and analyzing the facts and predictive outcomes, which can lead to better, more informed decisions. GIS technology supports critical thinking by enabling people to analyze each alternative and then come together to reach a mutual conclusion for the benefit of the community.

To help solve problems such as balancing land development and protecting the environment, a GIS could be applied to ensure that concerns over carbon footprints are included in the decision-making process. A developer wanting to maximize a site could point toward jobs and tax revenues as the positive outcome. A GIS application, on the other hand, can show storm-water runoff problems associated with minimal landscaping, the population loads of streets based on high-density land uses, the impacts of congestion and carbon loads, and the long-term costs of mitigating maximized development. In the end, GIS technology can expedite the decision-making process by staying on course, minimizing distractions, and instilling decision makers with the confidence that their reasoning was firmly based on the facts.

GIS helps Fort Bend County update emergency management

CHAPTER: Making decisions under pressure
ORGANIZATION: Fort Bend County, Texas
LOCATION: Rosenberg, Texas
CONTACT: Robert LaBarbera, GIS coordinator
rlb@co.fort-bend.tx.us
PROJECT: Emergency management data update
SOFTWARE: ArcGIS, ArcExplorer, ArcIMS, ArcView
ROI: Cost and time savings; increased efficiency, accuracy, and productivity

By Robert LaBarbera

Fort Bend County is one of the premier residential and commercial counties within the Houston metropolitan area. For more than fifteen years, Fort Bend has been in the top twenty counties in the United States for economic excellence, quality of life, living standards, and population growth. Excellent schools, affordable housing, and extensive recreational facilities have attracted families with impressive demographic profiles. These amenities can be found within the communities of Sugar Land, Missouri City, and many other master-planned communities within Fort Bend County. The broad appeal to live in Fort Bend County has fostered population growth, resulting in diversity of cultures, languages, and business types. World-renowned companies such as Fluor Daniel, Schlumberger, Sprint, and Texas Instruments have established headquarters in Fort Bend County seeking lower overhead costs and a better quality of life for their employees.

After the terrorist attacks of September 11, 2001, the federal government began promoting a proactive policy regarding terrorism. All large metropolitan suburban areas, including the Fort Bend County Office of Emergency Management (OEM), were under scrutiny for their ability to protect themselves from a terrorist attack.

Fort Bend County officials concluded that the OEM was not prepared for a local threat in its current administrative and structural form; the OEM needed upgraded hardware and software along with personnel who could manage the new technology. A revamped OEM would be needed to better protect the lives of Fort Bend County residents.

Fort Bend County is home to three major chemical plants, an electrical power plant, and minor light industrial complexes along with other plants and industrial complexes scattered throughout rural areas. Fort Bend County had an immediate need to be prepared to handle a possible chemical, biological, or nuclear attack or other form of terrorist threat. The county also needed OEM personnel to accomplish this task.

Fort Bend County officials hired Jeff Braun to direct the OEM and implement the new changes. Braun contacted the Fort Bend County GIS coordinator to learn more about GIS technology and how it could assist the OEM in the event of a terrorist attack. A GIS was first established in the Fort Bend County Engineer's Office in the mid-1990s and since then has steadily grown in capacity. In January 2004, the installation of an ArcIMS application made GIS available to any county department with Internet access.

Assessing OEM needs

With the help of the GIS coordinator, Braun identified several ways in which GIS could improve the emergency response capabilities of the OEM. The first need was for digital imagery to provide fire, police, health services, and emergency-response directors with the ability to pinpoint where an attack could occur. During an event, fire, police, and emergency response directors, possibly miles away from the scene of the event, would be able to see the incident and assess the situation within the confines and safety of the OEM building. This data would provide the ability to locate objects that ground personnel could not see because of massive smoke, fire, or debris; instruct ground personnel on the best routes either to or away from the incident; and quickly assess where to set up staging areas for residents in the event of a disease outbreak or massive contamination event.

There was also a need to pinpoint locations of terrorist attacks in real time and attach multiple forms of information to the locations such as the time of the incident, address, and the phone number of contact personnel. Another requirement was for real-time GIS map images to be projected so that personnel including telephone operators could view and report on the situation. It was essential for each PC in the operation command room to be GIS-enabled with the capability to stream textual communication between department directors and incident record-keeping personnel. The ability to provide printed copies of incident maps that were created digitally with the GIS in real time was important. Finally, sharing digital map layer information with local, state, and federal agencies was critical to emergency response efforts.

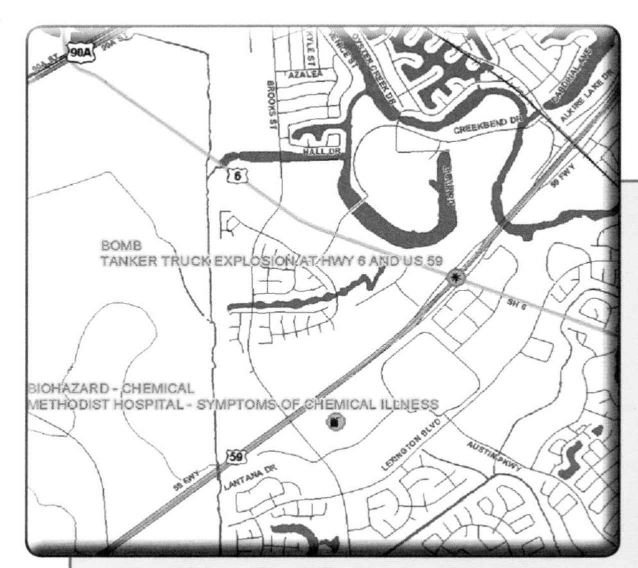

Figure 4.1 To better protect the public in the event of a terrorist attack, the Fort Bend County Office of Emergency Management had to be able to pinpoint locations of attacks and attach attributes to the sites.

The ability to provide printed copies of incident maps that were created digitally with the GIS in real time was important.

GIS solution for the OEM

The GIS coordinator had already built individual map layers. All that was needed was a way of transferring the data to the OEM. The coordinator proposed a plan to keep the OEM's GIS information up-to-date with the capability to create new layers in the event of an emergency or terrorist attack. The plan included:

- Establishing an FTP server in the Management Information System (MIS) department
- Providing bimonthly updates of the geodatabase to the OEM
- Providing the OEM with access to digital imagery and the Internet
- Providing the OEM with ArcView software for viewing the feature datasets within the geodatabase and building new layers on the fly
- Purchasing LCD projectors for displaying live GIS maps to the Operation Command Center and other locations

Practice exercise conducted

In May 2004, the county performed a practice terrorist drill to assess how well the current OEM could handle a terrorist attack, at what point the county would determine a need for federal intervention, and if that contact was appropriate for the circumstances. This process would point out deficiencies within the current configuration of the OEM. Texas A&M University's National Emergency Response and Rescue Training Center (NEERTC) conducted the drill in cooperation with the Texas Department of Public Safety and the Division of Emergency Management.

The drill involved all heads of critical departments, including road and bridge, health and human services, sheriff, fire, EMS, engineering, GIS, and purchasing/auditing. All charter cities within the county that employed their own fire, police, and EMS personnel also participated. Each city acted as a contact point and communication relay station for the county and initially handled the simulated attacks within city limits. If situations worsened, and the city's resources were depleted, then the city called on the county for additional resources.

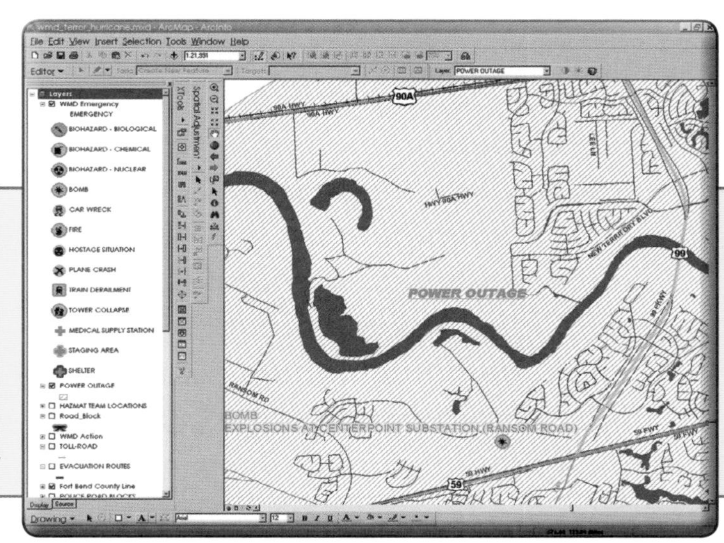

Figure 4.2 The county performed a practice terrorist drill involving all critical departments in May 2004 to assess how well OEM could respond to an attack.

The county acted as a central data-exchange point and assisted the cities up to the full commitment of their resources. The cities continued to provide information to the county about the status and progress of events. If incidents depleted county resources and personnel rapidly, the county alerted the federal government for assistance from the National Guard or other federal emergency or law enforcement agency.

On the first day of the drill, the GIS exceeded expectations. Incidents came in from numerous actors in the county and were mapped, with data attached to each incident. Aerial photography showed police, fire, and EMS locations that were hardest hit. Plume diagrams were drawn for county EMS personnel to better assess where certain segments of the population could be harmed by noxious chemical clouds.

The drill progressed with no major problems, and the county was successful in assessing the situation and making decisions under pressure—specifically, when to contact the federal government for assistance. Some changes were made to the OEM to better handle a terrorist event, including installing workstations for all departments heads, upgrading to ArcGIS software, and implementing emergency management software that enables department heads to view incidents in real time. More video screens and/or LCD projectors now enable all personnel, especially those in the command center and phone operator rooms, to view maps and incidents. A centralized software application connects cities with the county OEM for better data exchange.

GIS has evolving role in county activities

In the past, the Fort Bend County OEM handled emergency incidents without the use of any GIS or computer-aided mapping service. With domestic terrorism a real threat, the Fort Bend OEM has prepared itself for a nonconventional emergency. Because of its ease of operation and its ability to store and retrieve geographic information that can be shared easily from one department to another, GIS will continue to be an integral part of how the OEM better protects the residents of Fort Bend County.

The Fort Bend GIS has made progress since its inception. It has grown to be an integral part of the county's day-to-day activities and assists numerous county departments, the private sector, and the public. It will continue to grow and evolve into a more powerful and useful tool that is used by many people and organizations that reside or interact within Fort Bend County.

On the first day of the drill, the GIS exceeded expectations. Incidents came in from numerous actors in the county and were mapped, with data attached to each incident.

GIS plays a key role in Miami flood mitigation

CHAPTER: Making decisions under pressure
ORGANIZATION: Miami-Dade Office of Emergency Management
LOCATION: Miami, Florida
CONTACT: Soheila Ajabshir, GIS coordinator
soheila.ajabshir@miamidade.gov
Frank Reddish, emergency-management coordinator
PROJECT: Expediting a flood-control project
SOFTWARE: ArcInfo, ArcView
ROI: Cost and time savings; increased efficiency, accuracy, and productivity

By Soheila Ajabshir and Frank Reddish

South Florida flooding—it's a known hazard. Five or six inches of rain will cause a flood somewhere in Miami-Dade County. More than twelve inches of rain will cause a flood everywhere in Miami-Dade County. Because the topography of south Florida provides virtually no slope to the land and the water table is only a few feet down, heavy rainwater tends to stay put and not run off, even though there are canals everywhere.

Tropical Storms Gordon (1994) and Leslie (2000) along with Hurricane Irene (1999) each brought more than twelve inches of rain and massive flooding to the county. After Tropical Storm Leslie, Miami-Dade County's Office of Emergency Management (OEM) began the implementation of a flood-control program designed to end the flooding—a huge job that called for the cooperation and involvement of many federal, state, and local agencies.

Federal agencies included the Federal Emergency Management Agency (FEMA), the U.S. Army Corps of Engineers, U.S. Environmental Protection Agency, National Park Service, U.S. Fish and Wildlife Service, Bureau of Indian Affairs, and National Marine Fisheries Service. At the state level, the South Florida Water Management District, Florida Department of Community Affairs, and Florida Department of Environmental Protection engaged in the effort.

On the county level, the Miami-Dade County Office of Emergency Management, Public Works, Department of Environmental Resources Management, and the Board of County Commissioners Flood Management Task Force participated. The cities of Miami, Sweetwater, and West Miami were also involved. Two large engineering companies from the private sector, URS Corporation and PBS&J, participated. And other organizations such as the Miccosukee Tribe of Indians, the Audubon Society, and the Sierra Club also contributed.

A flood committee made up of representatives of these agencies immediately went to work to define the problem and recommend a solution. The C–4 Basin Initiative, named for the Tamiami Canal Basin, or C–4 Basin, was the result and entailed a massive engineering undertaking involving hydrology, hydraulics, and civil engineering. With all the agencies working together, the project was completed in four years. In addition to interagency cooperation, GIS played an integral role in the project's implementation and enabled the group to present the concept in a way that was easily visualized by the widely differing organizations.

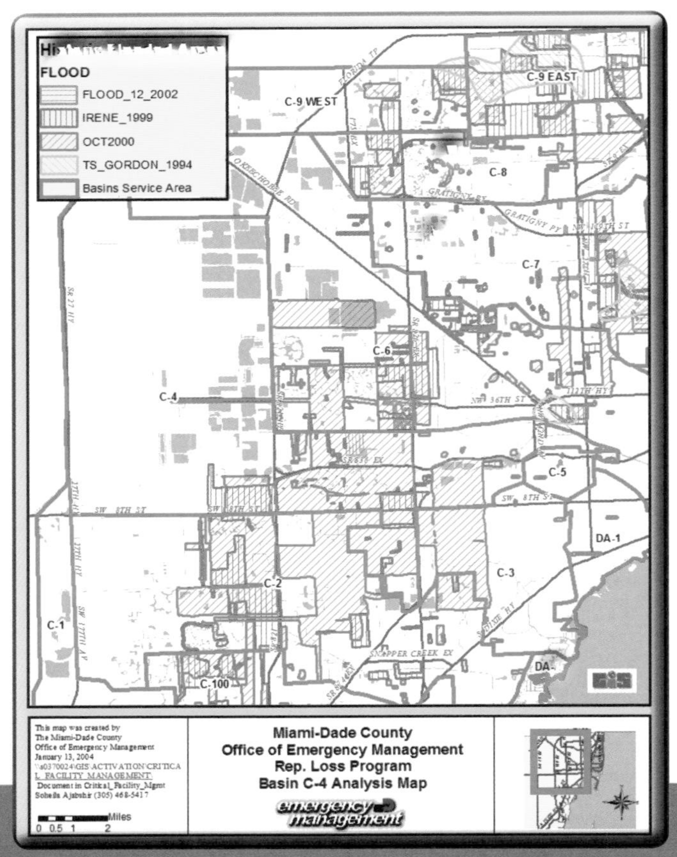

Figure 4.3 Map shows the areas in Miami-Dade County flooded in several severe tropical storms, including Irene and Gordon.

ESRI's ArcInfo and ArcView GIS software played a crucial role in the research phase. There are approximately twenty hydrological basins or watersheds within Miami-Dade County. ArcInfo software and land-elevation data were used to create the Canal's Basin Area Polygon shapefile, applying the contours of the various hydrological basins that indicate which canal or waterway forms each basin.

After these basin or watershed contours were identified, ArcView was used to "intersect" the basin layer with the population layer to estimate each basin population. The same geoprocessing capability of ArcView helped to determine the total number of lane-miles of flood-damaged roads, the number of flood claims filed with the National Flood Insurance Program, the number of loans issued by the U.S. Small Business Administration, and the number of assistance grants issued through FEMA's Individuals and Households Program. An inventory of FEMA repetitive-loss properties (properties where a flood insurance claim has been paid on the same structure two or more times) within each basin was also included.

In addition to interagency cooperation, GIS played an integral role in the project's implementation and enabled the group to present the concept in a way that was easily visualized by the widely differing organizations.

Using a mutually agreed upon weighting scale, the committee started the project with the C–4 Basin, which moves water from the Everglades in the west and flows into the Miami River (C–6 Canal) east of Miami International Airport. The Tamiami Canal starts in one national park (Everglades) and ends in another (Biscayne) and traverses an Indian reservation, a critical wetland, and several municipalities.

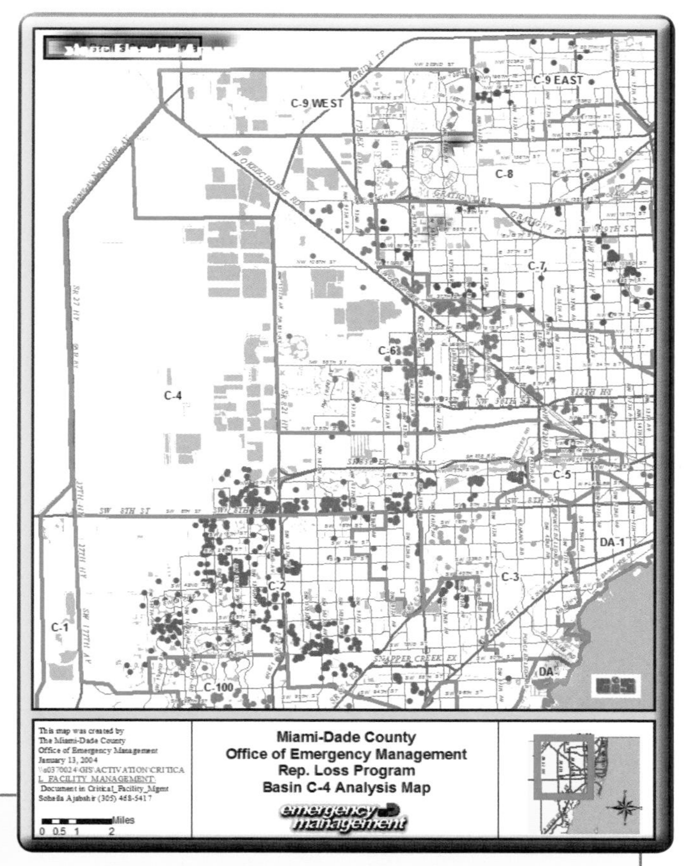

Figure 4.4 The Miami-Dade County flood-control program started with the C–4 Basin.

The timeline for the C–4 Basin Initiative was quite short compared with other public-works projects of this size. The flooding event happened in October 2000, with the concept for a solution formulated within two months. Engineering design and permitting began in January 2001, and by July 2001, the first-phase groundbreaking occurred for the installation of a massive pump to move water against the tide. The pump is large enough to fill a standard swimming pool in three seconds. The first phase of the project was completed in January 2002, less than eighteen months after the flooding occurred.

Another phase of the project was the installation of a similar pump in the adjacent Miami River to prevent the massive amount of water in the C–4 Canal from overwhelming the river's flow. The project also involved the construction of two 500-acre emergency detention reservoirs, including a supply canal and pumps to divert the water flow from the Everglades and create capacity in the C–4 Canal to handle the additional rainfall and runoff. Other phases included the installation of street drainage in the communities along the canal and the adding of a berm to raise the canal bank by several feet.

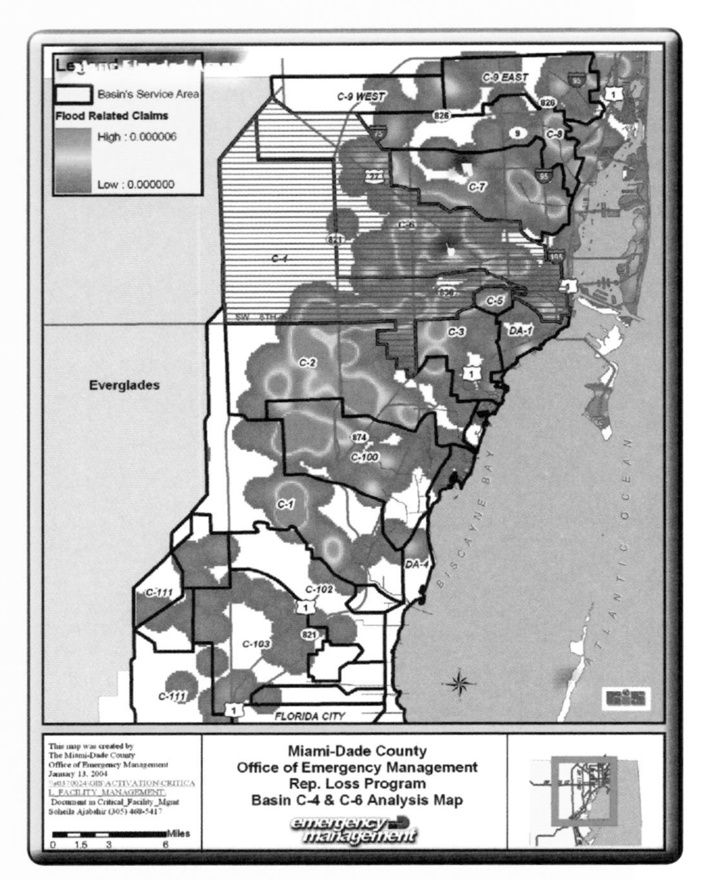

Figure 4.5 Map shows the historic flood concentration in the C–4 and C–6 basins based on repetitive-loss properties, with two or more claims.

All phases of the C–4 Basin Initiative were completed by early 2005. The first test of the project came in August 2005 when Hurricane Katrina passed through Miami-Dade County. To the relief of the flood committee and thousands of county residents, the C–4 Basin did not flood. Mitigation projects such as the C–4 Basin Initiative proved to be an effective method to better prepare communities for disasters. GIS proved it can be as valuable an asset in disaster mitigation as it is in disaster response. Because of the C–4 Basin Initiative, local, state, and federal governments as well as the residents of the basin have saved millions of dollars, and they are savings that will continue.

ArcInfo was used to create a polygon layer called Basin including each canal's service area. The geoprocessing capability of ArcView was used to intersect the Basin polygon layer with essential data such as population, number of flood claims, dollar values associated with the claims, and affected roads. Using ArcView Charts made it easy to visualize and compare the twenty basins and quickly prioritize the affected canals. The ArcView Spatial Analyst extension helped to show the density and concentration of the problem basins.

ArcInfo and ArcView software and extensions were used to create and then map a hydrological-basins table, which prioritizes the various basins and provides a starting point for the C–4 Basin Initiative.

Hydrological basins in order of priority								
Rank	**Basin**	**Population (30%)**	**Points**	**Flood claims (50%)**	**Claim points**	**Lane-miles damaged (20%)**	**Lane points**	**Total points**
1	C–4	493,377	42	5,312	124	50.60	49	215
2	C–6	750,197	64	3,885	91	44.50	43	198
3	C–7	714,843	61	4,908	115	4.20	4	179
4	C–8	178,988	15	2,028	47	32.40	31	94
5	C–9–E	141,982	12	2,248	52	8.40	8	73
6	C–2	188,440	16	820	19	29.20	28	63
7	C–3	443,625	38	78	2	.70	1	40
8	C–100	179,894	15	710	17	.40	0	32
9	C–1	90,000	8	560	13	4.30	4	25
10	C–102	40,800	3	718	17	4.82	5	25
11	Coast	164,528	14	0	0	10.00	10	24
12	C–103	68,650	6	157	4	12.70	12	22
13	C–9–W	32,000	3	1	0	2.50	2	5
14	C–111	35,331	3	0	0	2.20	2	5

Table 4.1 **Prioritization of flood risk for canal basins in Miami-Dade County.**

GIS software helped staff members save tremendous amounts of time as it geospatially analyzed all critical data and recommended priorities within a short time frame. It also helped policy makers visualize the priorities.

GIS proved it can be as valuable an asset in disaster mitigation as it is in disaster response. Because of the C–4 Basin Initiative, local, state, and federal governments as well as the residents of the basin have saved millions of dollars, and they are savings that will continue.

Rapid mapping analysis helps disaster victims

CHAPTER: Making decisions under pressure
ORGANIZATION: Peterson Air Force Base
LOCATION: Colorado Springs, Colorado
CONTACT: Tara Burkey, senior spectral/GIS analyst
tara.burkey@smdc-cs.army.mil
PROJECT: Posthurricane assistance
SOFTWARE: ArcView
ROI: Increased efficiency, accuracy, and productivity; more efficient allocation of resources; improved access to information

By Tara Burkey

To assist recovery efforts in the aftermath of Hurricane Katrina, the U.S. Army Space and Missile Defense Command (SMDC)/U.S. Army Forces Strategic Command (ARSTRAT), Measurement and Signature Intelligence (MASINT)/Advanced Geospatial Intelligence (AGI) Node (SMDC/ARSTRAT MASINT/AGI Node), located at Peterson Air Force Base in Colorado, was tasked by U.S. Northern Command (NORTHCOM) to provide potential helicopter landing zone (HLZ) maps that reflected the posthurricane conditions along the Mississippi coast.

To complete this task, the SMDC/ARSTRAT MASINT/AGI Node integrated advanced spectral analysis of commercial satellite imagery with GIS analysis of supplemental datasets, such as elevation and land cover, and produced a map delineating potential HLZs that met specific criteria. This map was disseminated by NORTHCOM to various recovery teams, including the 82nd Airborne Division and the Mississippi National Guard.

Specific criteria needed to be met for a valid HLZ analysis:

- Certain land-cover classes must be excluded—urban/built-up areas, agricultural areas, forested or heavily vegetated areas, wetlands, and water. Those land-cover classes with the greatest HLZ potential—grassland and barren/sparsely vegetated areas—needed to be isolated.
- New data reflecting current flooded and saturated areas must be incorporated in the analysis.
- Slope must be less than 15 percent.
- Vertical obstructions, such as towers and power lines, must be mapped as avoidance areas.

To render an analysis that met these criteria, two software packages were used—ESRI's ArcView software for geodatabase development, analysis of raster and vector datasets, and final map production; and Research Systems Inc.'s Environment for Visualizing Images (ENVI) for spectral analysis of newly collected satellite imagery.

Meeting the first two criteria proved to be of utmost importance, since recovery efforts required a rapid yet accurate situational analysis of posthurricane conditions along the Mississippi coastline. The IKONOS satellite's four-meter resolution multispectral imagery (MSI) of the area was collected by Space Imaging LLC and delivered to the

MDC/ARSTRAT MASINT/AGI Node for analysis. Using ENVI, the imagery was isolated into its individual blue, green, red, and near-infrared (IR) bands. By using the ratio of the blue band to the near-IR band, areas along the coast that had been flooded and/or saturated were easily identifiable.

Using ArcView, this ratio image was used as a threshold and displayed so that only the flooded and saturated areas were delineated as a single thematic layer. To further refine this analysis, a supervised classification of the new imagery was generated in ENVI. Land-cover classes reflecting up-to-date ground conditions were then incorporated as a layer in ArcView for use in the HLZ analysis. Supplementary land-classification information was provided by EarthSat GeoCover data, both to check the validity of the new classification and to fill in areas farther inland not covered by the IKONOS image swath. Finally, the land-cover classes were separated into "exclusion" and "potential HLZ" categories based on the land-cover types specified in the criteria.

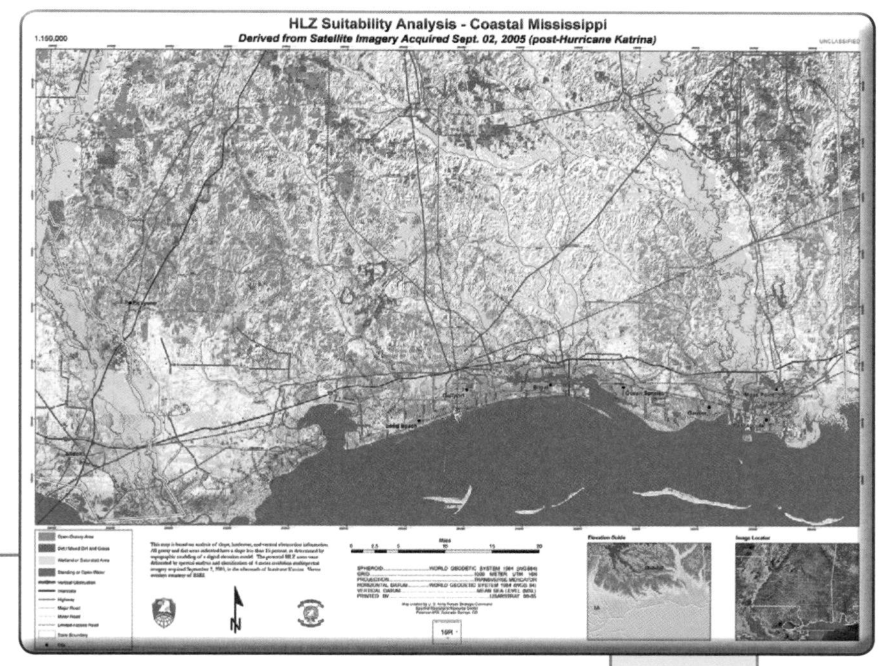

Figure 4.6 Helicopter landing zone map shows where choppers could safely land during Hurricane Katrina rescue, recovery, and cleanup.

Figure 4.7 Helicopters aiding in recovery efforts were provided with the latest, most up-to-date information on ground conditions through spectral and GIS analysis.

The next stage in the analysis required ArcView's Spatial Analyst extension. First, a digital terrain elevation model (DTED) was brought into ArcView and rendered in a slope file. Next, using Raster Calculator, only those potential HLZ land-cover classes that fell within areas where the slope was less than 15 percent were isolated as a layer. To further aid in terrain visualization, a hillshade was generated from the DTED for use as a background in the final map.

The final product that was delivered to NORTHCOM was fully created in ArcMap. The end goal was to provide the user a highly readable, current map to aid in locating HLZs during the recovery effort. To accomplish this goal, the data layers were stacked in the following manner: The hillshade was placed as a backdrop for topographic rendering, followed by the potential HLZ land classes that met the slope criteria, and then overlaid with street, city, and vertical-obstruction data.

Because of the methodology of integrating spectral and GIS analysis, the MDC/ARSTRAT MASINT/AGI Node was able to perform the analysis and produce a highly informative map reflecting posthurricane conditions within a matter of hours. This rapid turnaround enabled recovery teams to receive, disseminate, and use the map during the most crucial initial days of the recovery effort after Hurricane Katrina.

This rapid turnaround enabled recovery teams to receive, disseminate, and use the map during the most crucial initial days of the recovery effort in the aftermath of Hurricane Katrina.

Figure 4.8 GIS was put into action when Hurricane Katrina wreaked devastation across the landscape.

Exercise in decision support

Increasing immunization rates

Immunizations are cost effective public-health tools that have reduced the incidence of many diseases, such as polio, and eradicated others, including smallpox. Routine childhood vaccinations have dramatically reduced death and disability from illnesses such as measles, whooping cough, polio, and meningitis. Nevertheless, according to StatePublicHealth.org, more than 40,000 people die in the United States each year from diseases that are preventable with vaccinations or from complications resulting from those diseases.

State and local health departments have implemented Web-based immunization registries to inform health providers about the immunization status of their clients, remind families when an immunization is due, track adverse events, and increase immunization coverage rates. After Hurricane Katrina, many health providers were able to guarantee that displaced children received the vaccines they needed by accessing the Louisiana Immunization Registry.

National surveys provide helpful data about immunization coverage but can obscure pockets of need. These pockets of need are specific geographic areas within state or urban jurisdictions that contain large numbers of young children who are underimmunized and at risk for a vaccine-preventable disease. How can an immunization program manager efficiently identify pockets of need and then stimulate an increase in immunization rates with limited program resources?

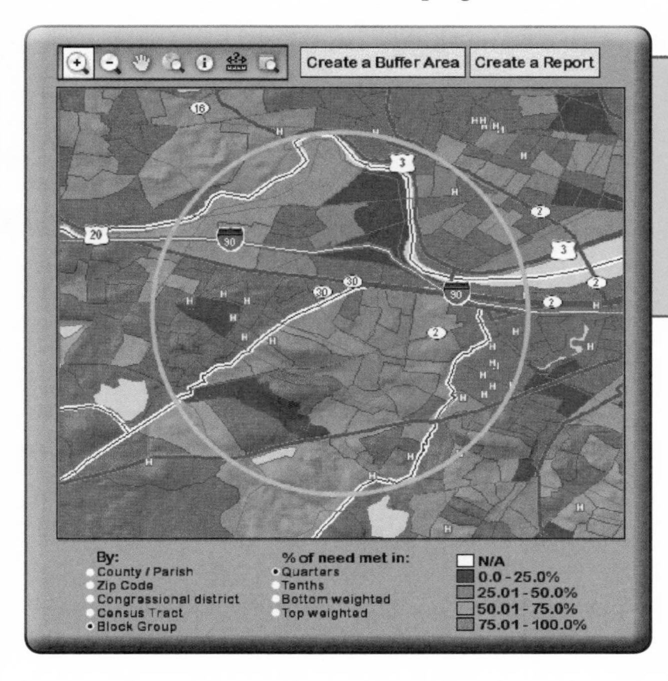

Figure 4.9 With GIS technology, an immunization program manager can easily visualize pockets of need (in red) in a region as well as potential providers (designated here with an "H"). Further analysis of the buffer zone (the area within the yellow circle) can help the manager identify priority locations for future immunization providers.

The GIS difference

Suppose an immunization program manager has information about pockets of need for a particular immunization (e.g., hepatitis B) during the past year. To target outreach efforts and prioritize limited program resources, the manager needs to analyze these pockets and compare them with a number of factors, such as the location of current immunization providers, location of potential providers, possible sites for outreach campaigns, and other demographic data.

Business decisions about where to begin can be difficult and time consuming when managers have only tables and paper maps to work with. However, with GIS software and the appropriate data, a program manager can quickly determine which private providers and public-health immunization clinics are in the pocket of need and analyze which ones are already participating in the registry. The program manager can also identify schools; clinics participating in the Special Supplemental Nutrition Program for Women, Infants, and Children program; churches; and municipal centers in the area to plan outreach and health-education activities.

Figure 4.10 Smallpox may no longer be a threat, but other illnesses such as measles and meningitis can be held at bay through regular childhood immunizations.

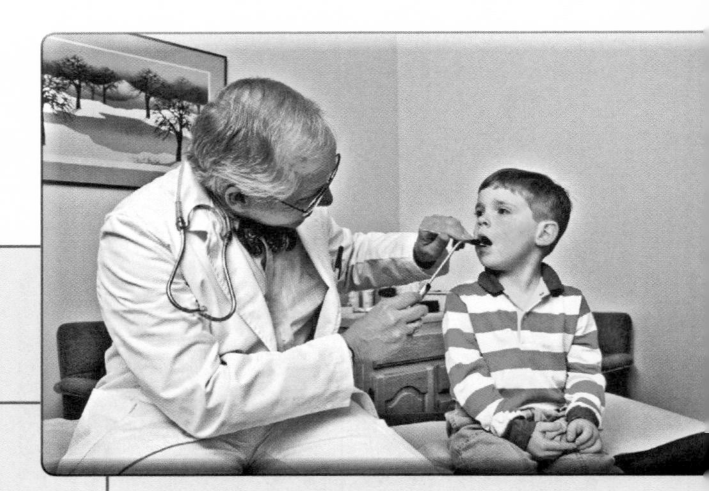

Figure 4.11 GIS can be used to create and monitor immunization registries to protect children's health—and the public's.

Using ArcGIS technology, the program manager can create an animation sequence of pocket-of-need maps to share with program staff and immunization providers. This helps build a shared understanding of the trends over time. The manager can also analyze areas of need in relation to demographic data to better understand the location of hard-to-reach populations. Because the risk of outbreaks of vaccine-preventable diseases is highest where concentrations of underimmunized children reside, health officials are encouraged to consider how using GIS technology can improve decision support in their immunization efforts

Decision support for allocating resources

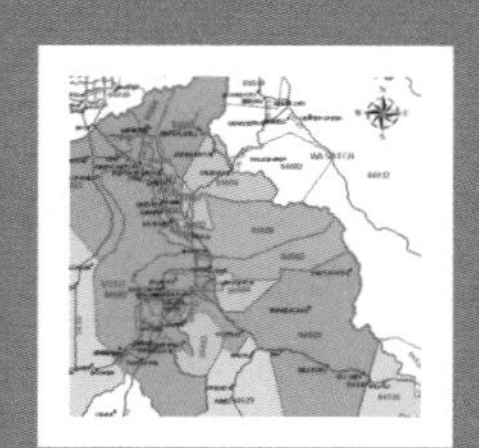
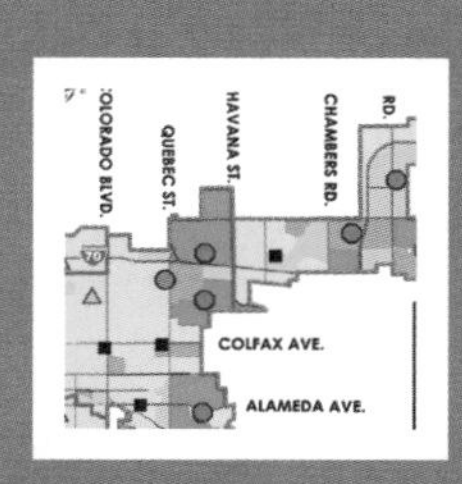

5

Governments have experienced increasing pressure to make savvy decisions with respect to resource allocation. Fair allocation of resources such as human skills, materials, inventory, financial resources, or information technology often has to be made amid budget constraints or with the "do more with less" philosophy handed down from management. With increased public scrutiny come the issues of performance measurement and accountability, increasing pressure on government leadership. Adding to the stress of managing resources, political pressures can weigh heavily on whether to deliver some services over others. These services can range from reducing the number of potholes to increasing public-safety personnel to supporting social-service programs.

Each of these decision constraints requires different analysis, with varying decision support tools and presentation methodologies. Most of the traditional tools available to policy makers have provided for modeling resource allocation based on individual tasks, but rarely do these systems allow for a holistic vantage point, or functional or cross-functional impact analysis after a decision is made. Governments have converged on using automated systems such as 311; enterprise resource planning (ERP); work-order, asset, facilities, and customer-relationship management (CRM); and executive dashboards. Can geography-based resource allocation offer an advantage over traditional methods or be combined with these systems for improved results?

Advocates of GIS technology work from the notion that all decisions have a geographic component. When a government executive needs to address the issue of increasing law-enforcement personnel, geography brings up certain questions: Where are the high-crime-rate areas? Where are high-risk populations? Are the crimes being committed within residential or business districts? Are there links between the time and day of the week and the location of the crime? Would the daytime and nighttime populations in a business district require the same or different allocation of police officers within designated police beats? The answers ultimately bring geography into the decision-making process and respond to performance measurement and accountability requirements.

Geography enhances tools such as capital-improvement viewers, because capital projects including parks, transportation, or utilities are geography based. A map can offer a realistic perspective of your decisions. Where are the capital projects with respect to the total number of people impacted? How much money is allocated to each project? Is the project on schedule? GIS technology offers a means by which executives can justify why certain activities are happening in a specific neighborhood. Taking the GIS-based capital-project allocation tool a step further, decision makers can look at capital projects in a certain location and look at cross-functional activities happening in the same location.

Have you ever wondered why a street continues to be dug up and paved over and over again? The problem lies in a lack of communication about the types of activities that are planned for a particular location. This inefficiency causes increases to material and staffing costs, not to mention the impacts of traffic delays on public-safety personnel, trash collection, and other government services. GIS technology can help managers integrate information from various sources and efficiently coordinate the activities of different parties working in the same location. For instance, a public electric utility could be planning an expansion while the sewer department has routine maintenance scheduled in the same area. GIS software can be used to coordinate these efforts and reallocate personnel, materials, and budgets for streamlined, cost-effective operations.

Denver Fire Department makes facility decisions based on GIS

CHAPTER: Decision support for allocating resources
ORGANIZATION: City and County of Denver
LOCATION: Denver, Colorado
CONTACT: Doug Genzer, senior GIS analyst
douglas.genzer@ci.denver.co.us
David Luhan, GIS supervisor
david.luhan@ci.denver.co.us
PROJECT: Fire Facilities Master Plan
SOFTWARE: ArcInfo, ArcView, ArcView Network Analyst, ArcView Spatial Analyst
ROI: Cost and time savings

By Doug Genzer and David Luhan

Located at the base of the towering and scenic Rocky Mountains, the city and county of Denver is the hub of the greater metropolitan area and home to more than 556,000 people, who generated 82,658 calls for service (CFS) to the Denver Fire Department's thirty-three fire stations in 2005.

In summer 2002, the city and county began developing a Fire Facilities Master Plan (FFMP) based on a comprehensive analysis of the fire department's existing facilities and performance. This master plan would become a road map for the future of the Denver Fire Department (DFD), and determine where new fire stations, as well as training and maintenance facilities would be built during the next twenty-five years.

GIS technology played a key role in this process and proved vital to performing analysis and displaying results, determining existing conditions, presenting proposed actions via live GIS demonstrations, and producing maps for the final document.

The plan was a collaborative effort involving representatives from Asset Management, the DFD, Technology Services–DenverGIS, the Budget and Management Office, the mayor's office, and the consulting firm Daniel Smith and Associates.

The project's scope

Technology Services–DenverGIS performed all of the geographic analyses for this plan. DenverGIS administers the development, management, and distribution of the city and county's geographic information. It provides standards, project management, data support and analysis, application development, training, and GIS services to city agencies and provides GIS products to the public.

The plan's goals were to

- Continue superior service by maintaining Denver's Insurance Services Office (ISO) Class 2 rating, which saves residents money on homeowner's insurance
- Provide equitable services throughout the city and county
- Meet the National Fire Protection Association response-time performance objectives of four minutes' travel time for the first arriving unit 90 percent of the time
- Improve response performance in developed areas
- Plan for adequate response in rapidly growing areas

- Establish a long-term plan that would set the parameters for all capital improvements and assure that monies are "spent once, spent right"

Coordination between many city and county agencies as well as outside entities was crucial in identifying and compiling requirements and developing a scope of work. The FFMP team consisted of one GIS analyst, one senior real-property agent, one technical writer, one budget analyst, and several fire department representatives. All analyses were performed on Microsoft Windows NT-based workstations.

The next part of the process included numerous meetings aimed at obtaining data, developing pertinent analyses and procedures, and educating involved parties on the viability of various alternatives for the location or relocation of fire stations.

Population, employment, and future CFS growth

The first questions that need to be answered when planning for the location of new fire stations are basic: What are the existing conditions, and in which areas are population and employment growth expected to occur? The Denver Regional Council of Governments generates population and employment projections by small geographical units called traffic analysis zones that are similar to census-block groups. GIS layers were created that

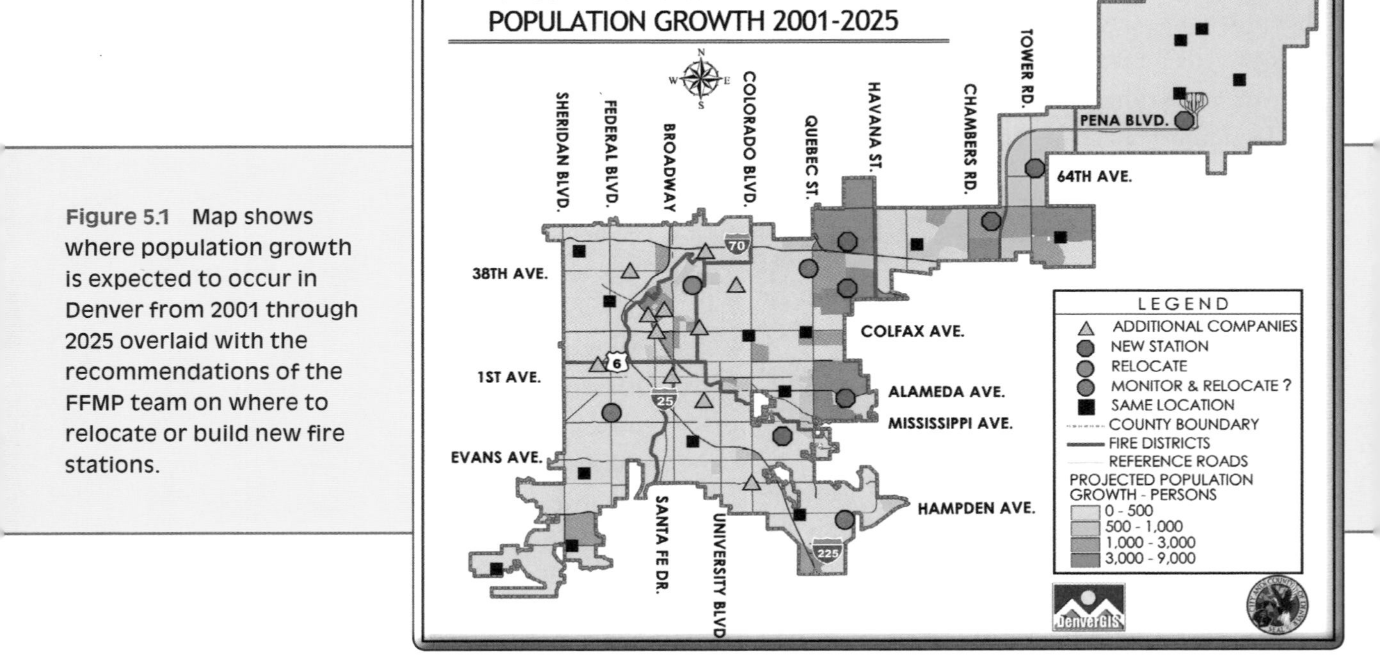

Figure 5.1 Map shows where population growth is expected to occur in Denver from 2001 through 2025 overlaid with the recommendations of the FFMP team on where to relocate or build new fire stations.

showed the predicted net growth in population in Denver between 2001 and 2025.

People generate calls for service, and so population is a major factor in determining the need for more fire stations. An area may have low nighttime population but a large number of daytime employees. Employment is a good indicator of daytime population. Additional map layers were developed that showed the predicted net change or number of additional workers in Denver between 2001 and 2025.

By looking at historic CFS data, the FFMP team determined that, on average, fifty-nine calls for service are generated per one thousand residents or employees. By applying this formula to the Denver Regional Council of Government's population and employment data, it was possible to project future CFS data. Layers detailing projected increases in calls for service through 2025 were created.

Response-time analysis

The DFD provided DenverGIS with a database containing the 2001 CFS. A GIS analyst used ArcView to geocode (digitally map) the CFS using the street address field. The analyst was able to geocode 90 percent of the 61,297 calls for service in 2001; standards issues with the data meant the rest of the records were unusable. Spatial Analyst was used to generate a density grid from the CFS point layer. Density grids show the distribution of a quantity per unit of space and are used to find geographic

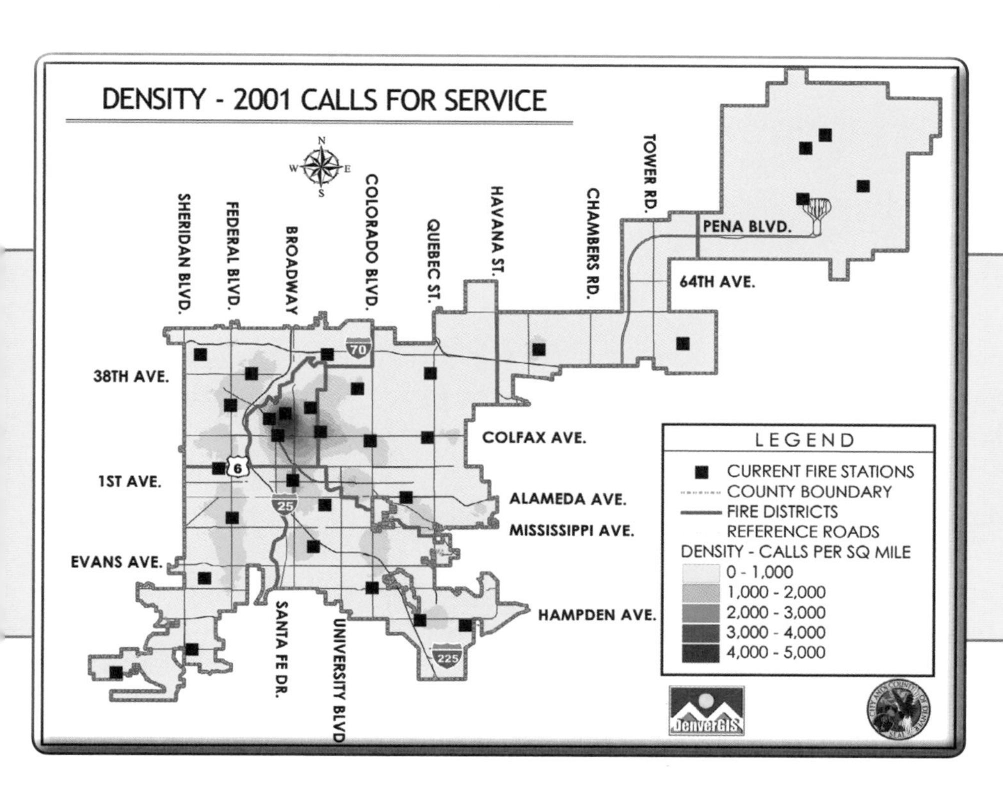

Figure 5.2 Density map indicates locations with the greatest call volume, a deterrent to the DFD meeting its travel-time goals.

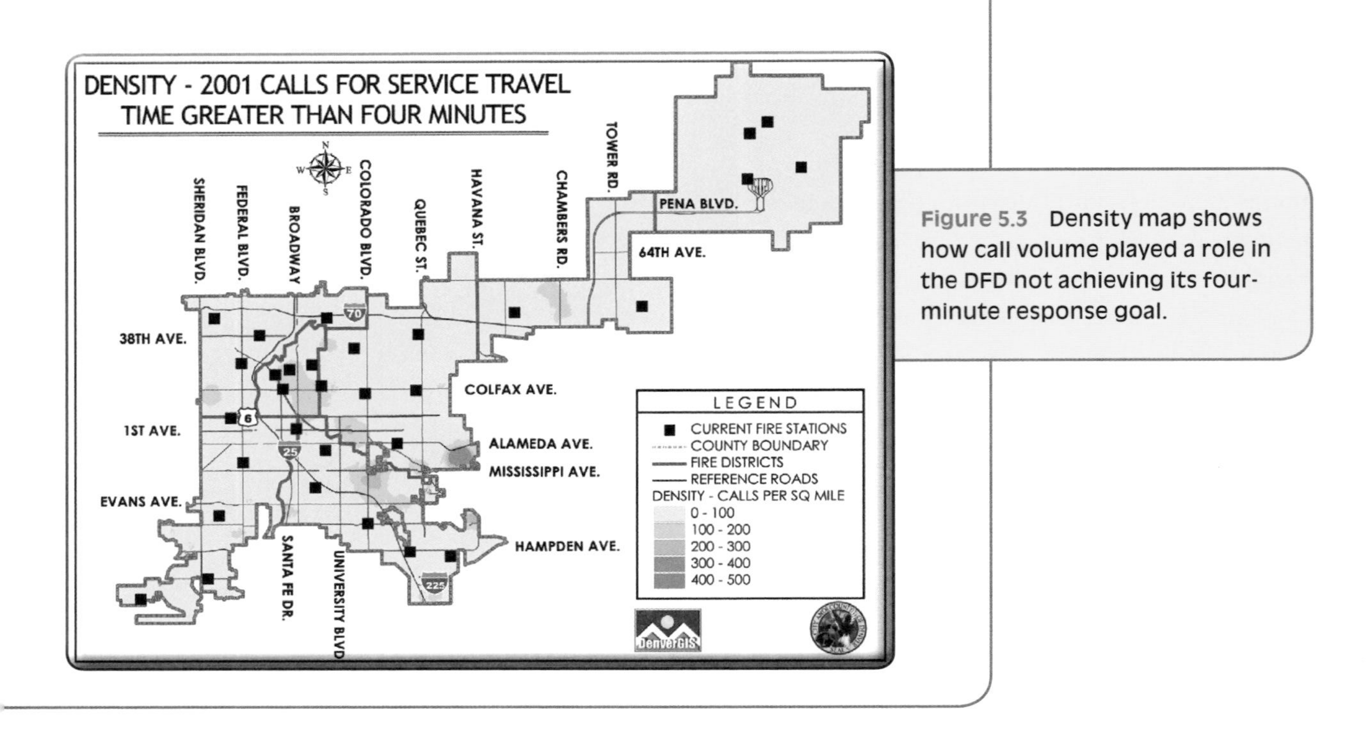

Figure 5.3 Density map shows how call volume played a role in the DFD not achieving its four-minute response goal.

patterns in data. This GIS layer showed the number of total CFS per square mile and would be crucial later in determining where call volume was an issue in meeting response-time goals.

Travel-time data was calculated, extracted, and joined to the CFS point layer. A density grid of CFS with a response time of greater than four minutes per square mile was created, which highlighted the areas where the DFD was not meeting its response goals.

To determine why the DFD was not reaching these areas, a network service-area analysis was run via ArcView Network Analyst using fire station locations and the city's street network. This service-area analysis used speed limits, one-way street designations, and road-closure information to determine how far a fire engine could travel from any fire station in four minutes.

Traffic congestion was another factor taken into consideration. A speed-reduction factor was calculated from traffic data provided by the Denver Regional Council of Governments and was applied to speed limits to replicate morning, post-rush-hour traffic congestion. The FFMP team decided that this would be a realistic estimate of average daily traffic congestion. The speed-reduction factor was applied to the road network by intersecting the network layer with the traffic analysis zone layer and calculating a new field using ArcInfo software.

By overlaying the results of the service-area analysis on top of the density grid of CFS with a travel time greater than four minutes, policy makers were able to see a correlation in the street network between areas that the fire department could not reach in four minutes because of station locations and the areas where the fire department was not meeting its response goals. These areas would need new fire stations.

The FFMP team determined that although the service-area analysis showed that certain areas should be reached in four minutes from existing fire stations, high call volumes in those areas interfered with four-minute response times. The team suggested that additional companies be added to existing fire stations in these areas to better address high call volumes. A layer of fire stations that could add additional companies was created. The fire department could add a company in these areas and monitor response times to determine if a new company increased response performance.

Proposed actions

This analysis led to recommendations made by the FFMP team regarding the optimum locations for new and relocated facilities. GIS streamlined this process by allowing all of the analyses to be overlaid and taken into consideration when planning for each future fire station site. Additional service-area analyses were conducted for each proposed fire station location. These analyses led to the following proposed actions:

- Construct six new fire stations
- Relocate two fire stations
- Monitor two existing fire stations for possible future relocation
- Add a company at eleven stations in areas where call volume was determined to be an issue

Presentations were given to the budget office, mayor, City Council, and Denver Planning Board. By using live GIS demonstrations, policy makers were able to clearly see the methodology and reasoning behind each recommended new or relocated fire station. The use of GIS in these presentations overwhelmingly convinced policy makers to back a plan that would cost money despite a period of tight budgets.

Return on investment

The major benefits of the FFMP are that Denver's residents are safer and that a process was created that can be easily updated with new data to ensure changing conditions are recognized and that proposed fire stations will be located in the right place.

A major financial return on investment was realized by one of the FFMP team's findings. The City of Glendale is surrounded by Denver and maintained its own fire department. The analysis performed in the FFMP showed an area just south of Glendale that the DFD was not reaching within its four-minute response-time goal. Since the Glendale Fire Department already had a fire station in this area, a four-minute service area analysis was run to see if this location would help DFD response times. Rather than build a new fire station, Denver and Glendale decided to join forces to better serve area residents and save money for both

The use of GIS in these presentations overwhelmingly convinced policy makers to back a plan that would cost money despite a period of tight budgets.

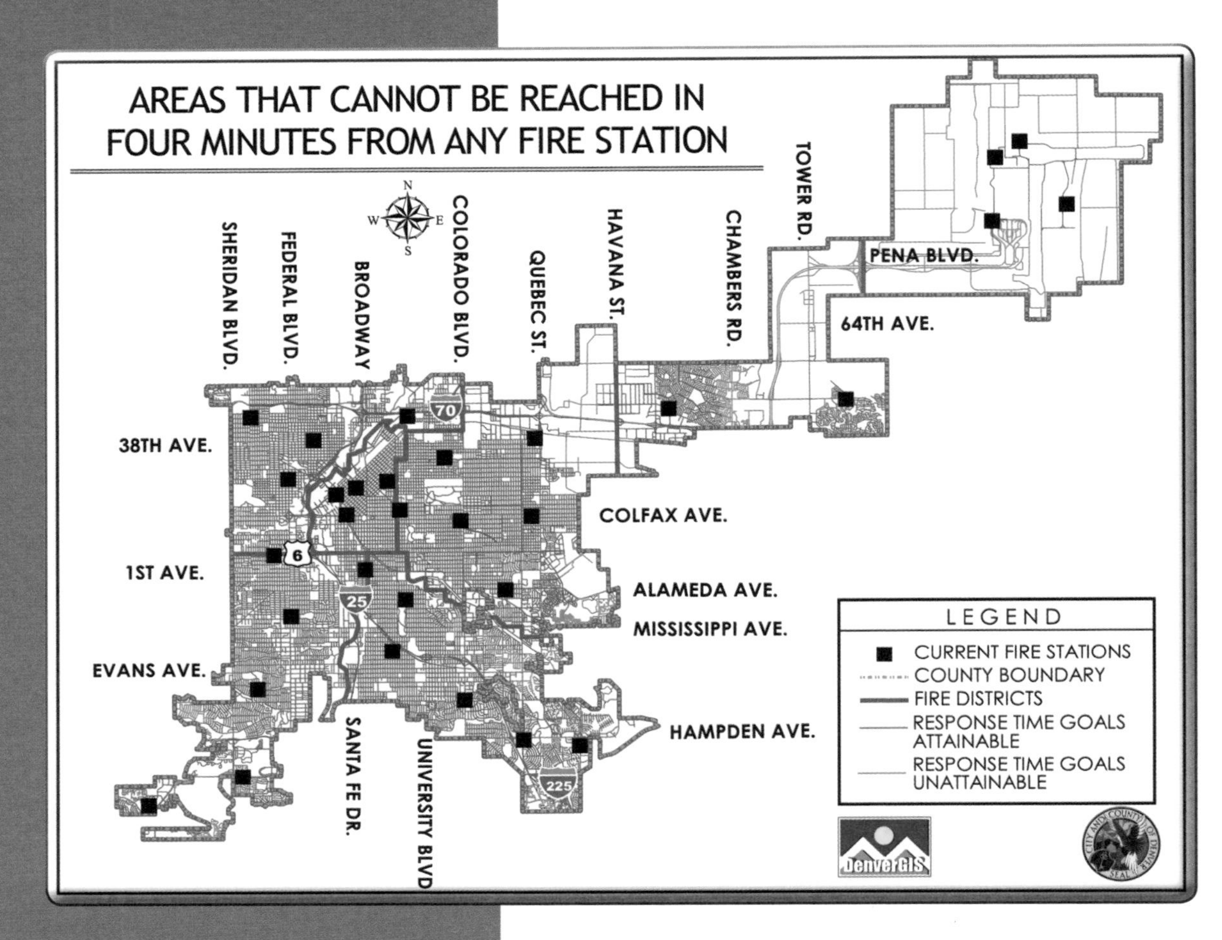

Figure 5.4 Road-network service-area analyses helped to determine where and why the Denver Fire Department was not meeting its goals.

Rather than build a new fire station, Denver and Glendale decided to join forces to better serve area residents and save money for both communities.

communities. The analysis and recommendations of the FFMP team directly led to the incorporation of the Glendale Fire Department into the DFD in 2005 for a savings of $2.5 million for a new fire station and $2 million for two fire vehicles. It also resulted in vastly improved delivery of services for both the City and County of Denver and the City of Glendale.

Another benefit can be found in increased efficiency and time savings for the GIS staff. All analyses performed were well documented, including the variables used. Recently, when certain analyses were rerun, what originally took more than one hundred hours of analyst time was completed in forty hours. Additional benefits were seen after using the improved DFD geocoding standards. In the most recent analysis, 99.6 percent of the DFD's CFS data was able to be located, up from the 90 percent that was usable in the initial analysis.

Since the FFMP was completed, one fire station has been built, one of the relocated stations is under construction, and land is being pursued for another. By knowing where to locate fire stations, a cost savings is realized in obtaining land before areas are fully developed and real estate values rise. By using GIS to evaluate fire station locations, Denver has saved money, become more efficient, and made its citizens safer.

The analysis and recommendations of the FFMP team directly led to ... a savings of $2.5 million for a new fire station and $2 million for two fire vehicles.

Pasadena visualizes and manages field-inspection activity

CHAPTER: Decision support for allocating resources
ORGANIZATION: City of Pasadena
LOCATION: Pasadena, California
CONTACT: John Reimers, project manager, Information Technology Support jreimers@cityofpasadena.net
PROJECT: Tracking code compliance
SOFTWARE: ArcIMS
ROI: Cost and time savings; increased efficiency, accuracy, and productivity; more efficient allocation of resources

By John Reimers

Best known for its New Year's Day festivities—the Tournament of Roses Parade and Rose Bowl game—Pasadena, California, is as proud of its rich cultural heritage and many historic properties as it is of its many modern residential amenities, vibrant commercial areas, and world-renowned institutions of higher education. Located in Los Angeles County, at the foot of the San Gabriel Mountains, and at the western edge of the San Gabriel Valley, Pasadena has a population of more than 136,000. Its borders encompass approximately twenty-three square miles and nearly 33,000 land parcels.

In 2000, the city began to develop a GIS primarily by a partnership between the departments of Planning and Development, Public Works, Water and Power, and Information Technology. Within Pasadena city government, an ArcIMS Web site is available to all city employees with intranet access. There are approximately twenty-five ArcGIS users in various departments throughout the city. The current GIS system and geodatabase is primarily driven by parcel-related data.

When Pasadena began developing its GIS, much of its tabular land-base data was stored in an Oracle database. This data is updated continuously through Accela Inc.'s Tidemark permitting and parcel-maintenance system. To avoid the cost and risk involved with updating two separate databases, the preferred scenario was for the GIS system to read the data, which was already replete with a wealth of parcel attributes. Recently, city staff launched a customized ArcIMS application (iMAP) that is linked to the parcel information tracking system. The goal was not only to be able to track code-compliance problems, but also to take proactive steps to minimize these issues. The permitting system is helping the Code Compliance section to do just that.

Using real- and near-real-time data, the iMAP application provides code-compliance inspectors and management with up-to-date information on housing density, status of code violations, vacant lots and vacant buildings, and land-use and parcel data. The code-compliance iMAP application enables multiple map views of code-violation issues

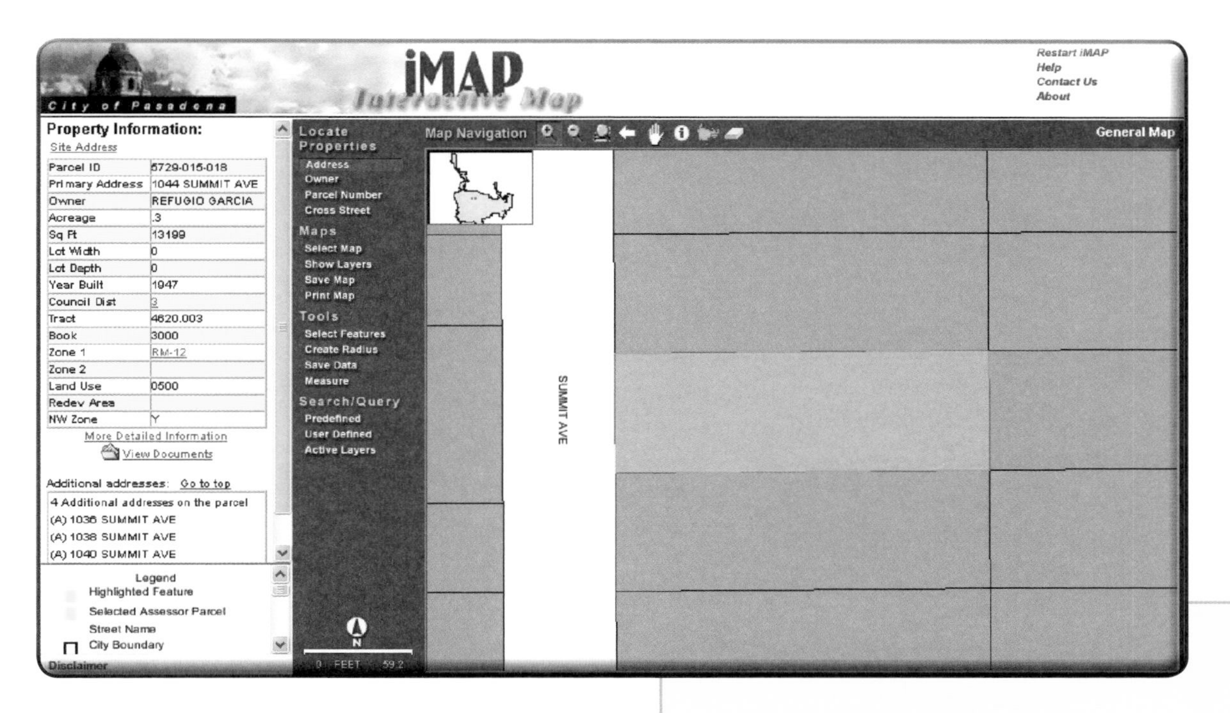

Figure 5.5 iMAP application provides up-to-date parcel information.

and potential trouble spots. These can be viewed at the parcel, neighborhood, or code-compliance territory level. Via iMAP, this information is available to all city management and staff with access to the enterprise network.

Available code-compliance map layers include the following:

- Schools
- Council districts
- Vacant buildings
- Vacant lots
- Residential-unit density (by parcel)
- National Register properties
- Landmark properties
- Landmark district
- Primary parcel addresses
- Assessor parcel lines
- Open compliance violations
- Violation severity
- Street names
- Proactive neighborhood status
- Code-compliance areas
- Land use
- Building footprints
- Condominiums
- Orthophotos

By turning the layers on and off, code-compliance staff is able to see in near real time where current problems exist. More importantly, code-compliance management is able to target compliance resources toward sensitive neighborhoods, including the following:

- More densely populated areas
- City cultural landmarks
- National Register and landmark neighborhoods
- Business districts
- Neighborhoods where violations are more prevalent

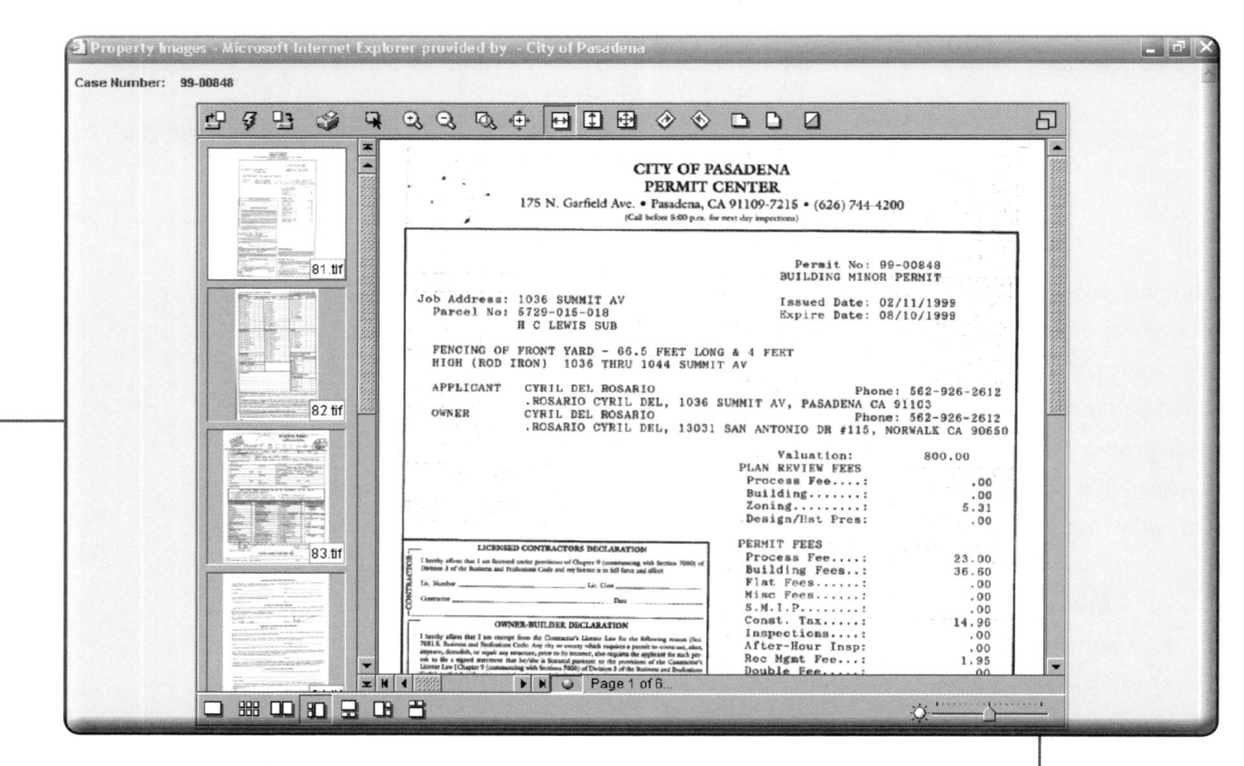

Figure 5.6 Using the iMAP application, staff can access and view images of permitting documents.

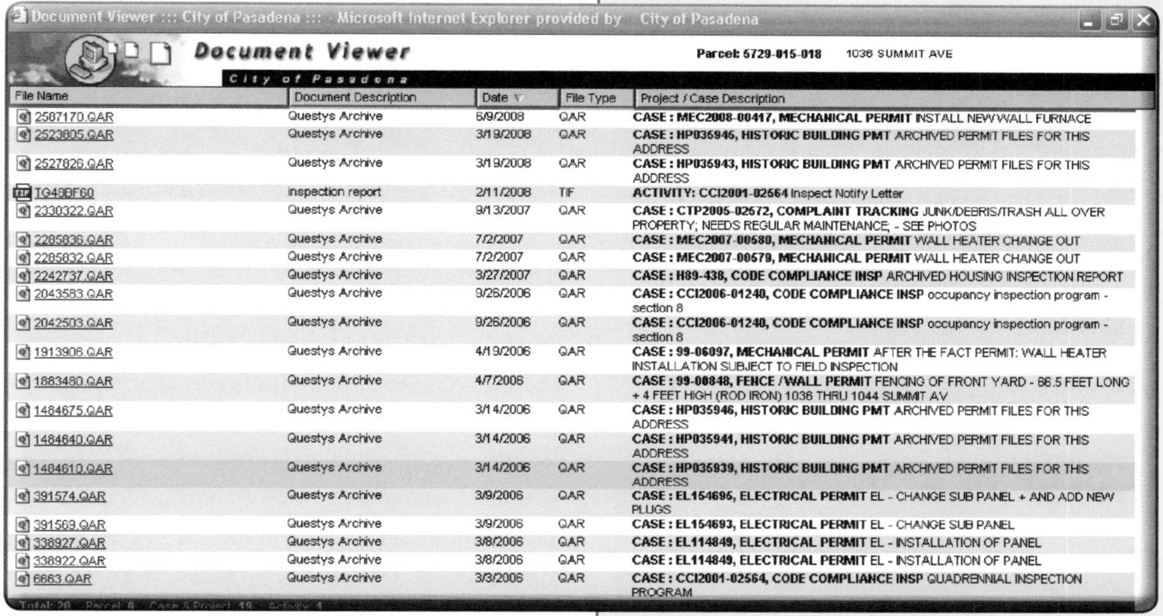

Document Viewer ::: City of Pasadena ::: - Microsoft Internet Explorer provided by City of Pasadena

Document Viewer

City of Pasadena

Parcel: 5729-015-018 1036 SUMMIT AVE

File Name	Document Description	Date	File Type	Project / Case Description
2587170.QAR	Questys Archive	6/9/2008	QAR	**CASE : MEC2008-00417, MECHANICAL PERMIT** INSTALL NEW WALL FURNACE
2523805.QAR	Questys Archive	3/19/2008	QAR	**CASE : HP035946, HISTORIC BUILDING PMT** ARCHIVED PERMIT FILES FOR THIS ADDRESS
2527826.QAR	Questys Archive	3/19/2008	QAR	**CASE : HP035943, HISTORIC BUILDING PMT** ARCHIVED PERMIT FILES FOR THIS ADDRESS
TG43BF60	inspection report	2/11/2008	TIF	**ACTIVITY: CCI2001-02564** Inspect Notify Letter
2330322.QAR	Questys Archive	9/13/2007	QAR	**CASE : CTP2005-02572, COMPLAINT TRACKING** JUNK/DEBRIS/TRASH ALL OVER PROPERTY; NEEDS REGULAR MAINTENANCE, - SEE PHOTOS
2285836.QAR	Questys Archive	7/2/2007	QAR	**CASE : MEC2007-00580, MECHANICAL PERMIT** WALL HEATER CHANGE OUT
2285832.QAR	Questys Archive	7/2/2007	QAR	**CASE : MEC2007-00579, MECHANICAL PERMIT** WALL HEATER CHANGE OUT
2242737.QAR	Questys Archive	3/27/2007	QAR	**CASE : H89-438, CODE COMPLIANCE INSP** ARCHIVED HOUSING INSPECTION REPORT
2043583.QAR	Questys Archive	9/26/2006	QAR	**CASE : CCI2006-01240, CODE COMPLIANCE INSP** occupancy inspection program - section 8
2042503.QAR	Questys Archive	9/26/2006	QAR	**CASE : CCI2006-01240, CODE COMPLIANCE INSP** occupancy inspection program - section 8
1913906.QAR	Questys Archive	4/19/2006	QAR	**CASE : 99-06097, MECHANICAL PERMIT** AFTER THE FACT PERMIT: WALL HEATER INSTALLATION SUBJECT TO FIELD INSPECTION
1883480.QAR	Questys Archive	4/7/2006	QAR	**CASE : 99-00848, FENCE /WALL PERMIT** FENCING OF FRONT YARD - 86.5 FEET LONG + 4 FEET HIGH (ROD IRON) 1036 THRU 1044 SUMMIT AV
1484675.QAR	Questys Archive	3/14/2006	QAR	**CASE : HP035946, HISTORIC BUILDING PMT** ARCHIVED PERMIT FILES FOR THIS ADDRESS
1484640.QAR	Questys Archive	3/14/2006	QAR	**CASE : HP035941, HISTORIC BUILDING PMT** ARCHIVED PERMIT FILES FOR THIS ADDRESS
1484610.QAR	Questys Archive	3/14/2006	QAR	**CASE : HP035939, HISTORIC BUILDING PMT** ARCHIVED PERMIT FILES FOR THIS ADDRESS
391574.QAR	Questys Archive	3/9/2006	QAR	**CASE : EL154696, ELECTRICAL PERMIT** EL - CHANGE SUB PANEL + AND ADD NEW PLUGS
391569.QAR	Questys Archive	3/9/2006	QAR	**CASE : EL154693, ELECTRICAL PERMIT** EL - CHANGE SUB PANEL
338927.QAR	Questys Archive	3/8/2006	QAR	**CASE : EL114849, ELECTRICAL PERMIT** EL - INSTALLATION OF PANEL
338922.QAR	Questys Archive	3/8/2006	QAR	**CASE : EL114849, ELECTRICAL PERMIT** EL - INSTALLATION OF PANEL
6663.QAR	Questys Archive	3/3/2006	QAR	**CASE : CCI2001-02564, CODE COMPLIANCE INSP** QUADRENNIAL INSPECTION PROGRAM

Total: 20 Parcel: 0 Case & Project: 19 Activity: 1

Figure 5.7 Document Viewer shows a list of documents associated with a specific parcel.

Using the iMAP application, staff can access parcel reports from the Tidemark information; display permitting and code-compliance violation history on a parcel; and view images of permits, other documents, and pictures associated with a property.

Since the application's implementation, city staff has added additional map services, data views, and functionality. Since the application's deployment in August 2006, the sixteen code-compliance officers who use it have been able to perform their duties more efficiently.

Assessing natural-hazards risks with GIS

CHAPTER: Decision support for allocating resources
ORGANIZATION: U.S. Geological Survey
LOCATION: Menlo Park, California
CONTACT: Laura Dinitz, operations research analyst
ldinitz@usgs.gov
PROJECT: Natural Hazard Risk Assessment
SOFTWARE: ArcMap
ROI: More efficient allocation of resources

By Laura Dinitz, Richard Champion, Anne Wein, Peter Ng, and Richard Bernknopf

Natural hazards pose significant threats to public safety and economic health worldwide. As people increasingly settle in locations that are exposed to natural hazards, financial losses from natural-hazard events have been rising and will continue to climb. State and local decision makers and leaders face the challenge of how to plan for and allocate scarce resources to invest in protecting their communities. A sound strategy to address this issue requires integrating knowledge and techniques from multiple fields, including geology, hydrology, geography, mathematics, statistics, and economics. A GIS provides the framework and technology in which these disciplines can be combined to inform these complex, inherently spatial problems.

In Squamish, British Columbia, Canada, U.S. Geological Survey (USGS) geography-discipline scientists worked with Geological Survey of Canada (GSC) scientists to apply the Land Use Portfolio Model (LUPM) decision-support system. The LUPM uses GIS methodology to integrate natural-hazard, land-use, damage, community asset, socioeconomic, and cost information to help communities evaluate alternative natural-hazard risk-reduction measures. The tool calculates, among other metrics, the expected value and uncertainty of the return on investment for a portfolio of natural-hazard mitigation investments. In this way, GIS technology serves as an invaluable strategic tool for decision making and policy development.

The model

The LUPM, developed by USGS geography-discipline scientists, is a GIS-based modeling, mapping, and risk-communication tool that can assist public agencies and communities in understanding and reducing their vulnerability to, and risk of, natural hazards. The tool supports decisions on community planning, mitigation measures, and allocating resources for protection against natural hazards. The model is inspired by financial-portfolio theory, a method for evaluating alternative, regional-scale investment possibilities based on estimated distributions of risk and return.

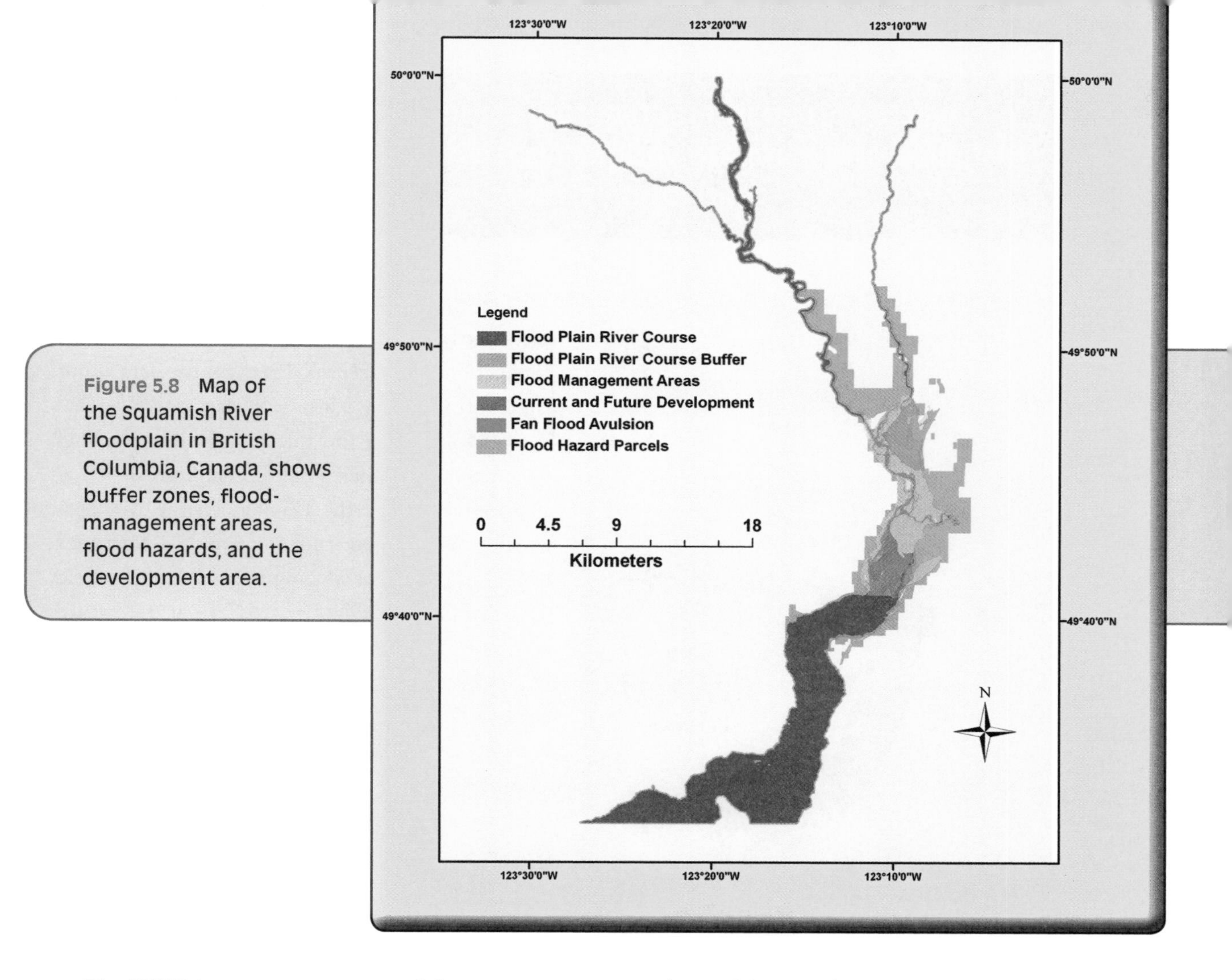

Figure 5.8 Map of the Squamish River floodplain in British Columbia, Canada, shows buffer zones, flood-management areas, flood hazards, and the development area.

The LUPM is an interactive tool for users to analyze, visualize, and compare various disaster-planning and loss-mitigation strategies. Required input data includes the probability of a hazard event; planning-time horizon; assets at risk (e.g., tax parcels); spatial conditional probabilities of damage, dollar value and/or vulnerability of each asset; and cost and effectiveness of risk-reduction measures being considered. The user selects a portfolio of locations and/or measures in which to invest a limited budget for loss mitigation. Then, for that portfolio, the LUPM calculates the total cost; number of locations mitigated; expected value and uncertainty of return on investment; expected loss; and expected value and uncertainty of community wealth retained. Finally, the user can examine maps that correspond to the results of each loss-mitigation policy, and compare and rank the policies according to their risks and returns.

The LUPM is an interactive tool for users to analyze, visualize, and compare various disaster-planning and loss-mitigation strategies.

A control library module for Microsoft .NET users, the LUPM can operate with Windows-based applications and access GIS layers for data input. The control library features a map interface that is used for editing and querying ESRI's ArcMap documents and other types of map files, as well as an interface called the Database Setup Manager, which is used for creating or opening a Microsoft Access database for storing hazard and scenario data. Additionally, data related to natural-hazard classes and events is maintained via a Hazard Events Manager interface, while data related to scenarios is handled via a Scenarios Manager interface. For each scenario, the LUPM generates a report summarizing the effects of a loss-mitigation portfolio on community wealth. The report, along with output for each parcel, can be saved to the database for future reference. The software is updated as needed to accommodate mathematical development, new GIS programming techniques, and other project needs.

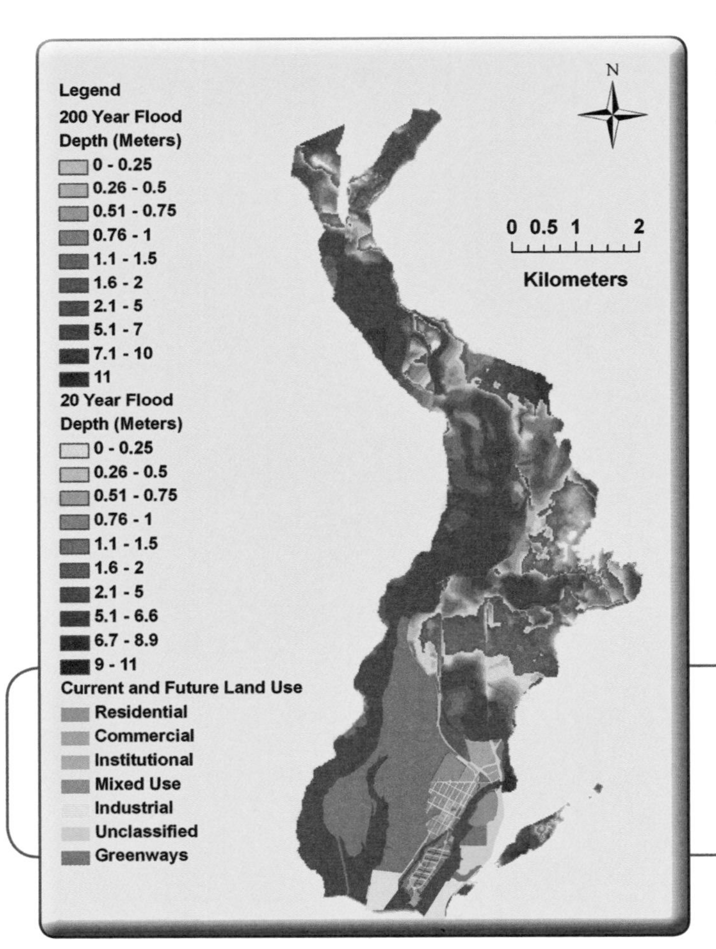

Figure 5.9 Flood depths for two types of flood events are shown for the Squamish River floodplain planning area, which includes both green space and development uses.

Role and benefits of a GIS in the LUPM

There are several benefits to implementing the LUPM within a GIS. First, the equations used to calculate event probabilities, loss, and risk are encapsulated behind a straightforward user interface, simplifying communication and enabling visualization of the predicted outcomes of alternative risk-reduction measures. Furthermore, the LUPM interface enables users to spatially select a portfolio of risk-reduction strategies.

Users can spatially display, at the parcel level, the progression of four steps in a hazard-event analysis, including hazard, damage, loss, and risk. Parcel-level analysis is important, because it is one scale of analysis on which decisions can be made. A GIS also enables users to visually compare the risk profiles of various scenarios (hazard magnitudes and/or proposed future concept plans). For communities that face multiple hazards, the hazard layers are analyzed within the GIS to determine which hazard dominates the risk, and to identify parcels affected by more than one hazard.

Using LUPM in Squamish, British Columbia

USGS and GSC scientists are collaborating to provide a scientific basis for growth management and natural-hazards mitigation along the Pacific coast boundary. The goal of this joint project is to incorporate indicators of natural-hazard risk into an evaluation of future-growth plans and to support investigation of the effectiveness, costs, and benefits of various risk-reduction strategies. In a pilot study, the USGS-GSC science team has collaborated with the district of Squamish in British Columbia, a coastal community about 40 miles north of Vancouver. The community will be the gateway to the 2010 Winter Olympics in Whistler, and the population is predicted to almost double in the next twenty-five years to approximately thirty thousand.

Squamish is exposed to at least three types of natural hazards—debris flows, earthquakes, and floods. Analysis reveals floods to be the most recent, frequent, and severe natural hazard to affect the development area. GIS is being used to integrate scientific and socioeconomic information to incorporate flood risk into the planning process. The immediate objective of the LUPM is to evaluate the potential losses and risk from flood events for recent land-use plans and eventually to support natural-hazard risk-reduction decisions.

For communities that face multiple hazards, the hazard layers are analyzed within the GIS to determine which hazard dominates the risk, and to identify parcels affected by more than one hazard.

A GIS analysis compared two potential flood events—a less-frequent two-hundred-year flood and a more-frequent twenty-year flood—to better understand the range of possible outcomes. The planning area included in the land-use plans consists of green space and a development area. The green space, designated for conservation and recreation, is subject to deeper flood depths; the development area designated for current and future construction, though still in the floodplain, is at a higher elevation. North of the planning area, the depth and extent of a two-hundred-year flood is significantly greater than that of a twenty-year flood. Within the planning area, however, these two floods are of approximately equal depth, except along the margin of green space. The differences appear to be least in the development area; however, flood-damage estimates (derived from functions of depth and velocity) indicate that building damage from the two-hundred-year flood is twice that of the twenty-year flood, although a twenty-year flood is ten times more likely to occur.

The LUPM was used to run scenarios to analyze the sensitivity of risk results to decision parameters. It was noted that Squamish's future chosen concept plan increased the risk, though less so than either doubling the planning horizon from twenty-five to fifty years or using property values (an upper bound on losses) instead of structural values. A hypothetical analysis of the cost-effectiveness of two mitigation strategies illustrated risk-return trade-offs between the two options.

There have been a number of preliminary successes in the efforts to apply the LUPM in Squamish. A 2005 GSC-USGS workshop presenting GIS analyses and LUPM results to Squamish community planners raised awareness and educated planners about how to integrate the risk of natural hazards into the planning process and the potential benefits of incorporating science into planning decisions. GIS-based LUPM risk analysis confirmed floods as the most important hazard

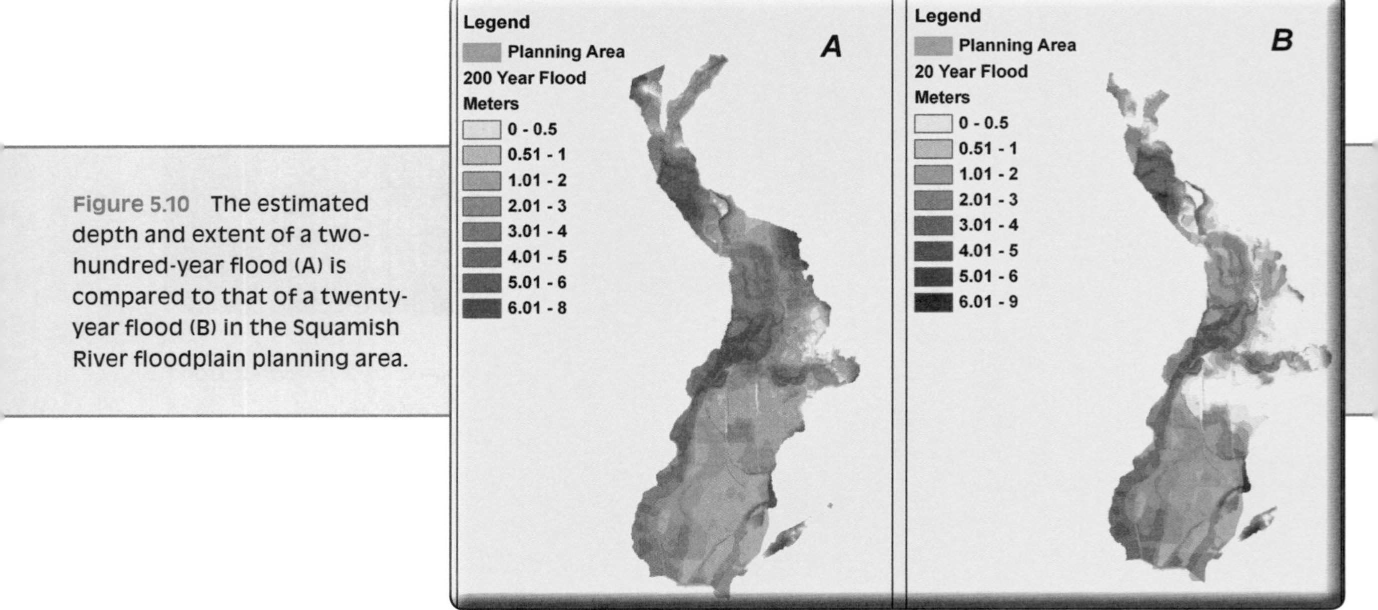

Figure 5.10 The estimated depth and extent of a two-hundred-year flood (A) is compared to that of a twenty-year flood (B) in the Squamish River floodplain planning area.

to guide risk-reduction efforts. Independently, the GSC undertook a multiple-hazard risk perception study that was compared with the modeled risk results. It revealed that overall, participants ranked the risk for three types of hazard (flood, earthquake, and debris flow) consistently with the model results but expressed relatively more concern for large earthquakes.

Acknowledgments

We thank our partners in the GSC, a branch of Natural Resources Canada, including Murray Journeay for his project vision and data compilation, Chang-jo Chung for his contribution to the probability calculations, and Sonia Talwar for her help in working with the District of Squamish.

Analysis reveals floods to be the most recent, frequent, and severe natural hazard to affect the development area. GIS is being used to integrate scientific and socioeconomic information to incorporate flood considerations into the planning process.

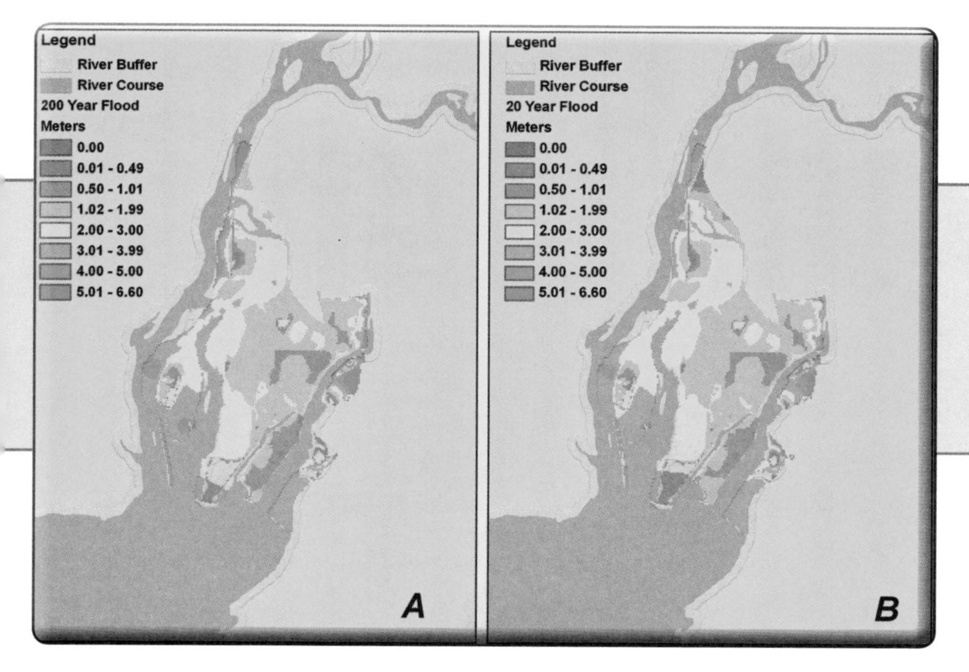

Figure 5.11 GIS is used in the Land Use Portfolio Model to target areas of lesser damage to development in the Squamish River floodplain for a two-hundred-year flood (A) and a twenty-year flood (B).

Utah uses GIS to map immunization registry data

CHAPTER: Decision support for allocating resources
ORGANIZATION: Utah Department of Health
LOCATION: Salt Lake City, Utah
CONTACT: Sandra Schulthies, data-operations manager for Utah Statewide Immunization Information System
sandys1@utah.gov
Yukiko Yoneoka, data-quality analyst for Utah Statewide Immunization Information System
PROJECT: Immunization assessment
SOFTWARE: ArcGIS
ROI: Increased efficiency, accuracy, and productivity; more efficient allocation of resources; improved access to information

By Sandra Schulthies and Yukiko Yoneoka

The Utah Statewide Immunization Information System (USIIS) at the Utah Department of Health is an immunization registry that stores and consolidates immunization records for residents of Utah. USIIS receives data from a variety of sources, including local health departments, private health-care providers, and community health centers. Data is submitted by ASCII flat files through a Web application. As of December 1, 2005, it was estimated that 34 percent of private providers, 100 percent of local health departments, and 58 percent of community health centers were submitting patient and immunization data. Authorized users can access USIIS to ensure adequate patient immunization.

The National Immunization Survey (NIS), a yearly assessment of immunization rates for children age 19 months to 35 months old conducted by the Centers for Disease Control and Prevention (CDC), showed that immunization rates in Utah declined significantly from 2003 to 2004. According to the report, in 2003, the up-to-date immunization rate for Utah children getting four doses of diphtheria, tetanus and pertussis (DTP) vaccine; three doses of polio vaccine; one dose of measles, mumps, and rubella (MMR) vaccine; three doses of hemophilus influenza Type B (Hib) vaccine; and three doses of hepatitis B (HepB) vaccine, was 80.4 percent. In 2004, Utah's immunization rate was reported to be 75.4 percent, showing a 5 percent decrease in the state's immunization rate.

The USIIS program developed a strategy to identify where the lowest immunization rates are in Utah by using the immunization registry, ArcGIS ArcView software, and digital maps provided by the Utah Automated Geographic Reference Center. The analyses were conducted for each of the twelve local health districts. Identifying the areas of lowest immunization rates in Utah helped determine where to focus efforts for increasing the number of adequately immunized children.

Children age 19 months to 35 months old as of December 31, 2004, were selected from the USIIS database. They were divided into two groups: those

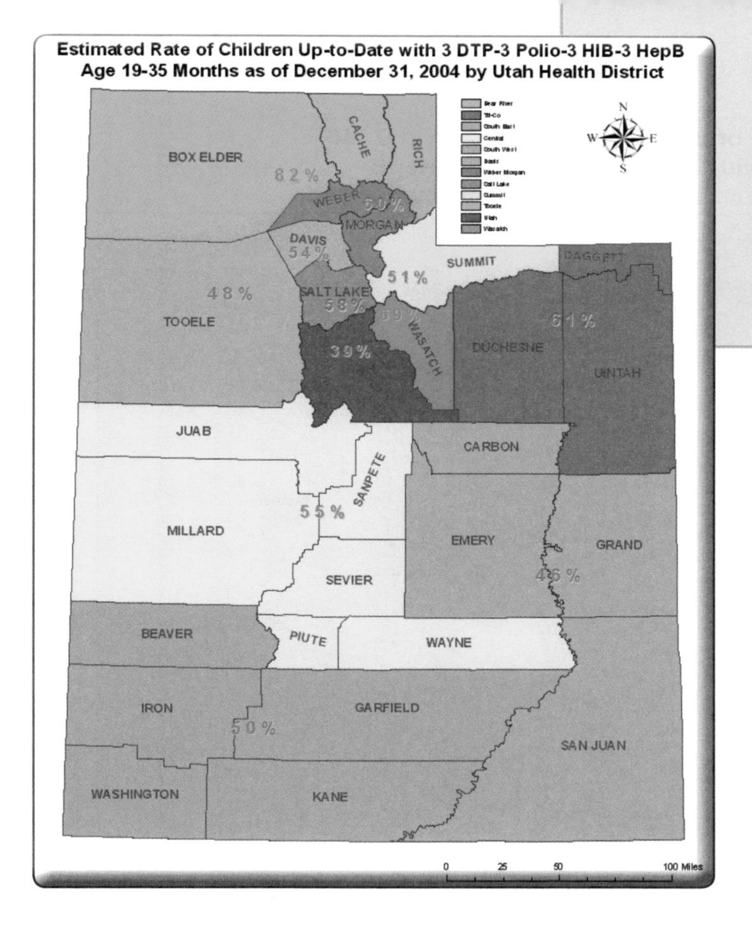

Figure 5.12 Map shows the disparity in estimated toddler immunization rates across Utah's twelve health districts.

whose immunizations were up-to-date with three doses of DTP, three doses of polio, three doses of Hib, and three doses of HepB vaccine, and those whose were not. Their physical addresses were geocoded by ZIP Code and placed on the Utah state map. The number of children was summed up within each health district to calculate the percentage of children whose immunizations were up-to-date versus not up-to-date. Estimated percentage intervals for adequately immunized children were determined by calculating 95 percent confidence intervals for proportions of up-to-date children.

Identifying the areas of lowest immunization rates in Utah helped determine where to focus efforts for increasing the number of adequately immunized children.

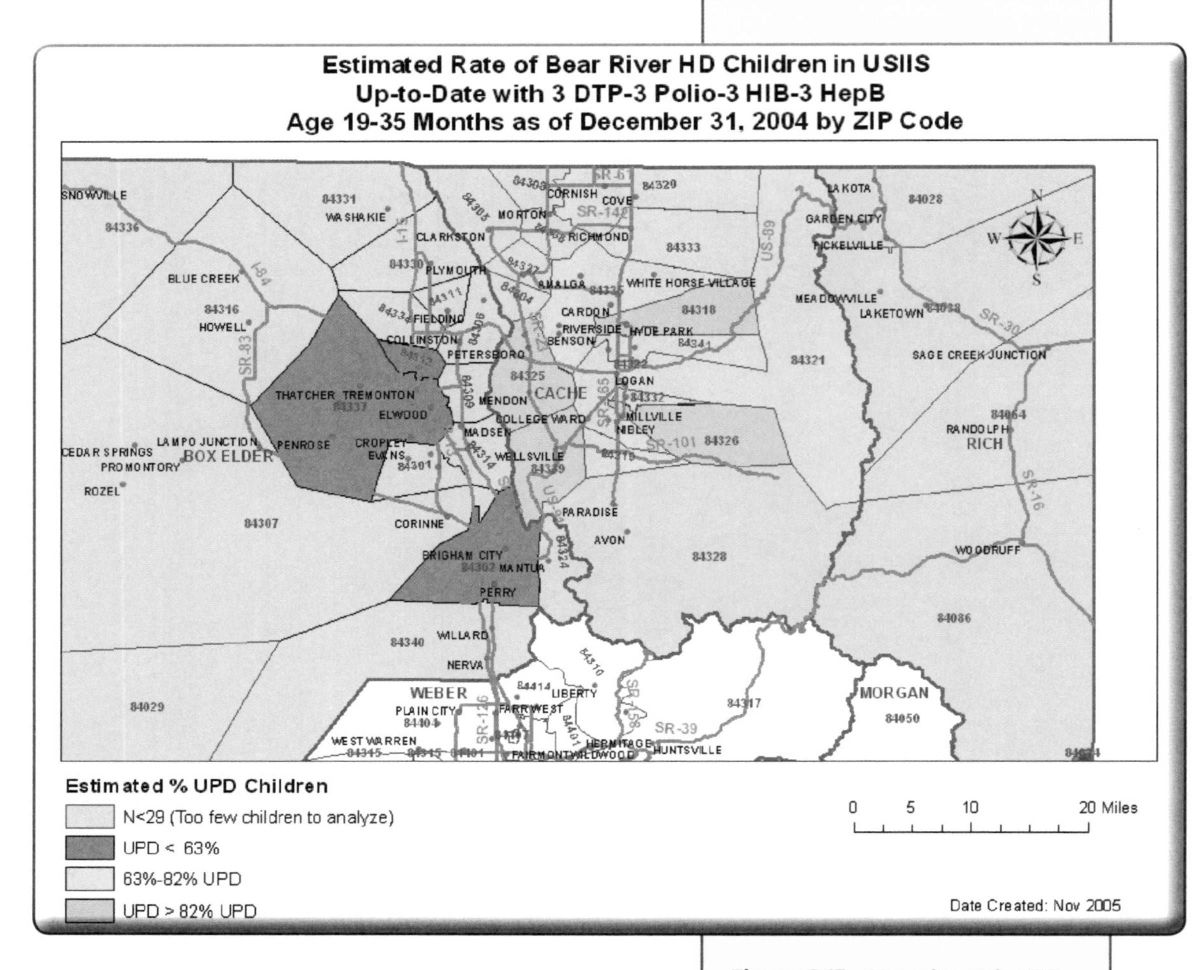

Figure 5.13 Map shows by ZIP Code how toddlers in Bear River Health District have the highest immunization rates.

Estimated rates of adequately immunized children plotted on the Utah Health District map gave a clear representation of data that was easy to compare. They ranged from 82 percent immunized in the Bear River Health District to 39 percent in the Utah County Health District. Thus, administrative decisions were made, without the cost of further investigation, to focus on uncovering and alleviating barriers to immunization in Utah County. The return on investment for those efforts will be realized when studies are completed and interventions are implemented.

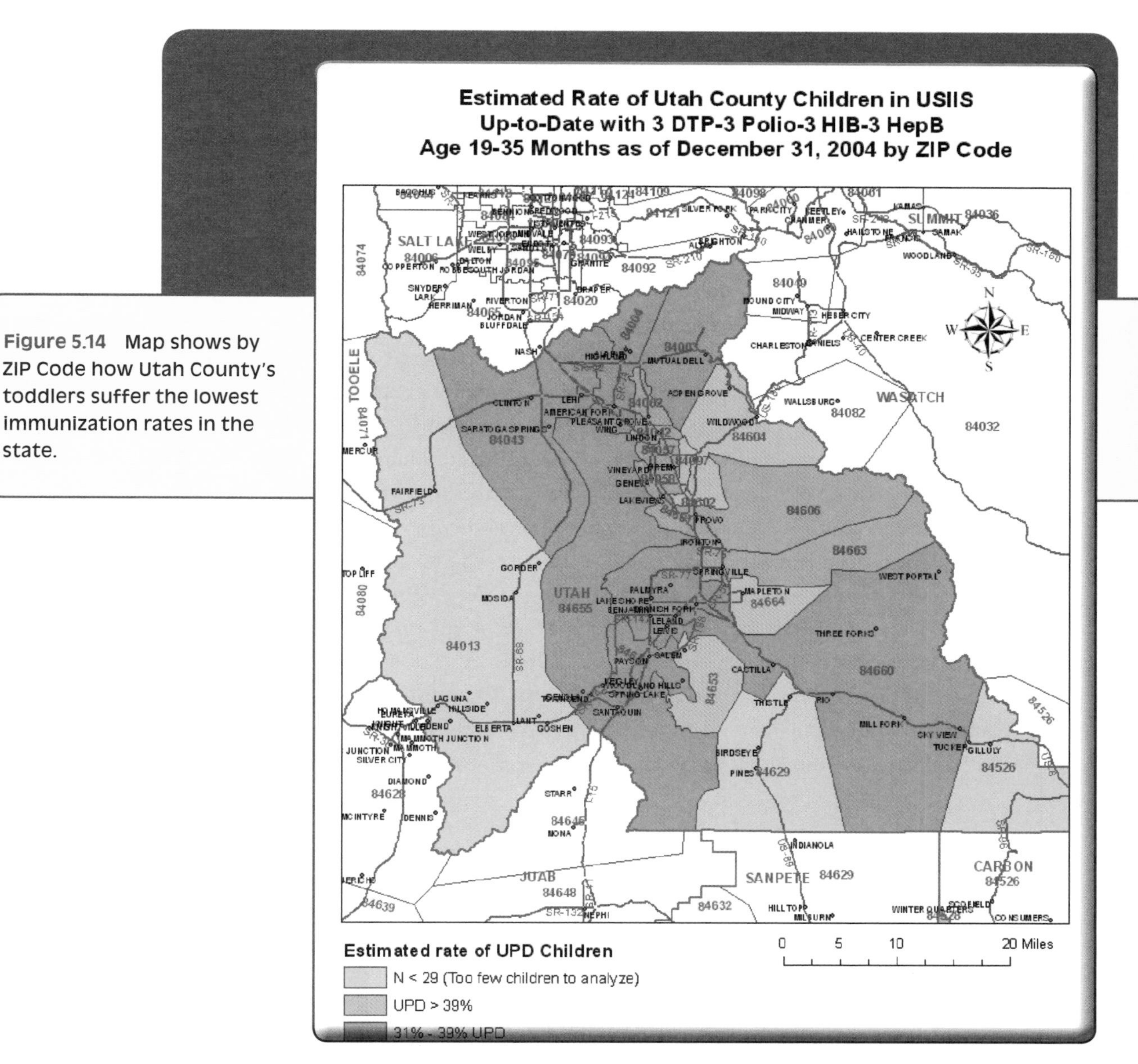

Figure 5.14 Map shows by ZIP Code how Utah County's toddlers suffer the lowest immunization rates in the state.

References

Morbidity and Mortality Weekly Report (U.S. CDC), July 30, 2004/53(29); 658–661, National, State, and Urban Area Vaccination Coverage Among Children Aged 19–35 Months. United States, 2003.

Morbidity and Mortality Weekly Report, July 29, 2005/54(29); 717–721, National, State, and Urban Area Vaccination Coverage Among Children Aged 19–35 Months. United States, 2004.

Exercise in decision support

How can the county clerk prepare for multiple elections in a year?

In the United States, county clerk's offices typically administer elections with two primary tasks: establishing voting-precinct boundaries and finding suitable locations for polling places.

To determine the shape and size of a precinct, staff members consider a variety of factors, including demographic data from census reports and physical boundaries, which can range from naturally occurring features such as rivers to constructed barriers such as freeways. Polling places must be located within precincts, and voters must be properly assigned to avoid overwhelming a site.

Often, polling places are located at facilities that serve other purposes, such as churches, schools, government offices, or even a private homeowner's garage. Election officials must ensure that structures can be properly utilized during polling hours as well as meet other requirements such as accommodating a specified number of voting machines. Certain accessibility issues, such as handicapped parking, proximity to public transit, and wheelchair ramps, are also taken into account.

In addition to determining precinct boundaries and polling locations, staff members need to find facilities suitable for training election workers as well as distribution points for election equipment and materials that will be routed to and from the polls. In large or densely populated counties, thousands of election workers are needed to staff polling places. To optimize efficiency, more than one distribution point for election equipment might be required. Add to this the fact that many counties must administer several elections a year, and it becomes clear that elections require year-round planning.

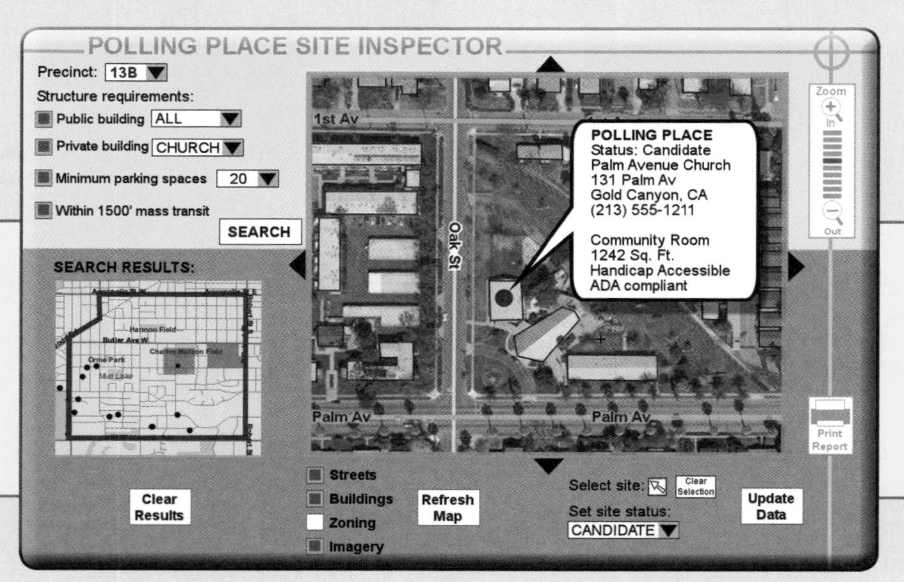

Figure 5.15 GIS users can quickly identify suitable polling locations based on whether they meet specific structure requirements.

The GIS difference

Figure 5.16 GIS is used to ensure that all voters can be accommodated.

To enhance geospatial problem solving, a county clerk's office often needs to become a more active customer of the county's GIS department. For example, much of the information required to create precinct boundaries or find suitable locations for polling places is stored in the county's GIS database. The GIS department performs analyses, creates analytical tools for decision making, and builds online applications such as polling-place locators that can help cut down on the volume of support calls during elections.

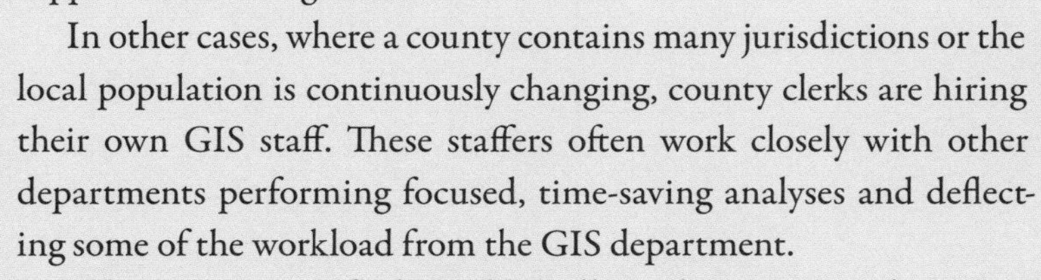

In other cases, where a county contains many jurisdictions or the local population is continuously changing, county clerks are hiring their own GIS staff. These staffers often work closely with other departments performing focused, time-saving analyses and deflecting some of the workload from the GIS department.

When it's time to find suitable polling places, GIS tools can significantly lessen field research. To learn about a site, an analyst selects from an inventory of prescreened candidates (e.g., schools, government facilities, and homes) and confines the search to a specified area, or precinct. From the selected set, the study is further refined using attributes of those structures, such as square footage and number of parking spaces, as well as proximity to bus stops and transit stations. The facilities are graded based on their level of suitability. High-resolution aerial imagery (stored in the GIS) of the candidate sites is viewed by inspectors before they visit the physical location so that they can familiarize themselves with site characteristics not easily seen from street level. Reviewing proposed sites before going into the field eliminates unqualified sites, empowers decision makers, and speeds up the decision-making process.

A simple change in the parameters of the tool enables the analyst to search for potential election-worker training centers, equipment distribution sites, or any other facility needed to support the election. GIS will also help election staff find efficient routes for delivery of voting equipment.

Making decisions on the fly

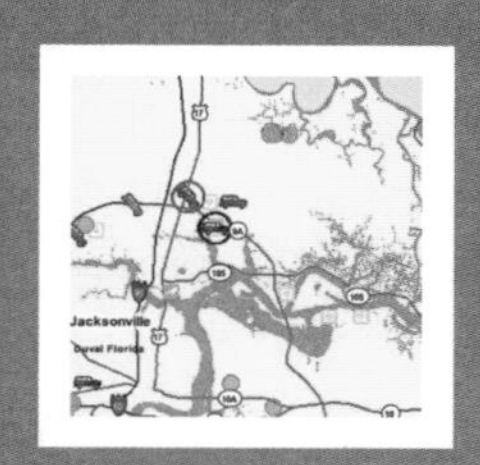

There's a time in every manager's career when decisions need to be made instantaneously—on the fly. These decisions often have to be made during a crisis such as a disaster, health scare, or financial or political unrest. They happen at a time when delaying a decision could escalate retaliatory responses, or simply when time is not on your side. These situations create an atmosphere in which all eyes and ears are focused on a single group or individual and when other managers are quick to relinquish authority to someone else. Making the wrong call can lead to a public-relations nightmare.

This type of judgment call strays from traditional processes where decisions are made after a healthy debate and all aspects of an issue are considered. Similarly, conventional reasoning via weighted options is usually not applied. Nor is there time for management to pinpoint good reasons for not following a course of action before committing to it. In the end, leaders look to trusted advisers to provide quick input and then make a decision based on intuition.

The question is, can GIS support the demands of making a decision on the spot? If our government leaders acknowledge that GIS is more than a means of outputting a paper map, the answer is yes—absolutely, especially when the power of GIS is embraced as a knowledge-management tool. GIS evolved from a need to present high-quality visualization of situational analysis in the form of an executive-information system, or "executive dashboard," which provides a common operating picture. These systems geographically reference all of the data related to a situation, whether it is financial or demographic, or whether it is ground truth. A GIS is the type of decision-support tool that can capitalize on the strengths of multiple disciplines and integrate them into a common interface.

A GIS can store history and models to be retrieved later and run instantaneously. For example, decision makers faced with flooding can quickly look at the current perimeter of rising waters, run an in-depth analysis based on similar events, assess the population affected, and move quickly to stage evacuation routes and emergency shelters. As the situation changes, government leaders can monitor the activity in real time via mobile data feeds and can ensure successful outcomes by taking corrective actions on an as-needed basis. This type of decision making shifts the accountability to those who better understand the mitigation procedures.

Finally, GIS can be used in on-the-fly decision making to communicate decisions via Web portals and media outlets in a simplified, easy-to-understand forum. A GIS can serve to reduce emotions through facts and solutions.

Those dependent on your decisions can observe the implications as the situation changes. GIS technology presents information that is readily accessible and understood, and it can be used to demonstrate that a community's leaders clearly understand the conditions and context of an event. Using GIS technology to communicate a decision can diffuse argument, debate, and doubt to offer a solution that everybody can agree on. It reinforces clarity of purpose, credibility of leadership, and the integrity of an organization.

Executive dashboards improve situational awareness

CHAPTER: Making decisions on the fly
ORGANIZATION: City of Jacksonville
LOCATION: Jacksonville, Florida
CONTACT: George Chakhtoura, GIS manager georgic@coj.net
PROJECT: Executive information system
SOFTWARE: ArcIMS
ROI: Increased efficiency, accuracy, and productivity; automated workflows; more efficient allocation of resources; improved access to information.

By George Chakhtoura

The City of Jacksonville, Florida, is a consolidated form of government that comprises Duval County, Florida, and four incorporated municipalities—Atlantic Beach, Neptune Beach, Jacksonville Beach, and Baldwin. With a population of approximately 850,000 and a 2 percent annual growth rate, Jacksonville employs nearly 7,000 people who render various services to an estimated three hundred thousand households and twenty-one thousand businesses.

Increasingly, GIS technology has become an integral part of Jacksonville city government. Integrating geospatial databases with tabular attributes has provided a new dimension to data analysis and information management. Whether it is incident reporting or tracking mobile assets, the software has begun to predominate in operations and is relied upon by managers for decision support.

With advances in computer programming and data management, developers have the ability to manipulate data and provide accurate and real-time feedback from live situations. This has provided Jacksonville with the ability to implement a valuable executive information system used to streamline operations, track live data, and disseminate information to managers for effective decision making and improved situational awareness.

Citizen Active Response Effort

The Citizen Active Response Effort (CARE) is a service-based program that enables Jacksonville residents to call in or report via e-mail issues, complaints, and requests for service. A clearinghouse organization receives each request and routes cases to appropriate departments. Requests are entered in a database and tracked as their status changes. Estimated completion times are governed by predetermined thresholds for particular case types.

Whether it is incident reporting or tracking mobile assets, the software has begun to predominate in operations and is relied upon by managers for decision support.

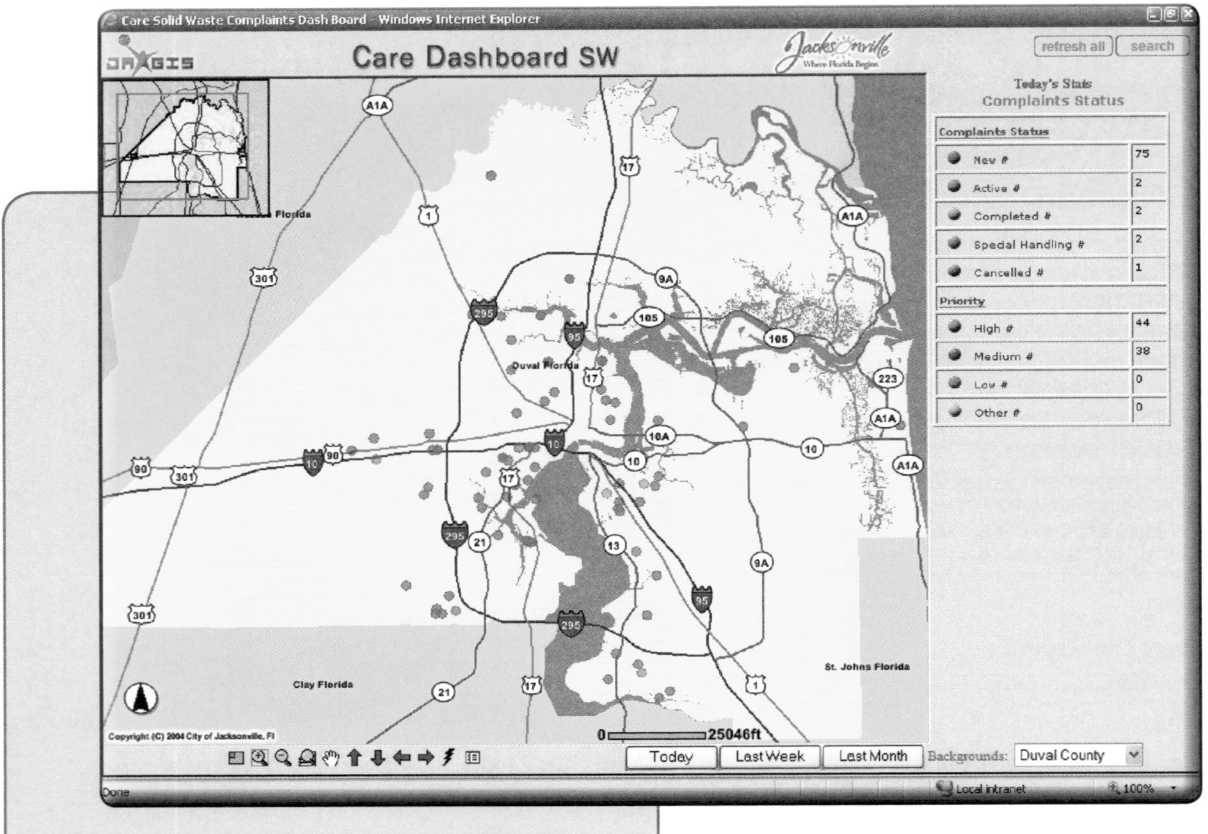

Figure 6.1 CARE dashboard for solid waste registers seventy-five new citizen complaints.

The CARE solid-waste application was developed to track cases related to solid-waste management. Daily, the agency receives nearly two hundred requests for service relating to trash, recycling, yard waste, and bulk-waste removal.

The CARE Dashboard is an ArcIMS application that provides a continuous update of cases in a live environment. The dashboard serves as a tool to keep the department head and other managers abreast of case status without having to manipulate the screen or use a mouse and keyboard. The application runs as a screen saver and enables managers to assess the situation in the field at a glance. The background consists of land-base layers for reference and orientation. Since its inception in May 2005, approximately seventy-three thousand cases have been processed with a 30 percent improvement in the level of service.

Debris management

The debris-management application Mapboard was developed to track the entire cycle of debris collection and disposal for posthurricane cleanup. The program complies with Federal Emergency Management Agency regulations and ensures appropriate tracking to yield eligibility for maximum reimbursement.

The goal and objective of the GIS-based executive Mapboard is to provide decision makers and elected officials with the ability to track the progress of posthurricane cleanup efforts from the emergency operations center or from anywhere within the county network. The viewer consists of a map of the county subdivided into a grid system reflective of the data collected. Colors and symbols are used to highlight progress and milestones reached. Statistics reflecting progress made and disposal-site capacity are also provided for planning purposes.

Auto Vehicle Locator

The Auto Vehicle Locator (AVL) project is aimed at tracking data from mobile assets in the field of operations. This system uses a proprietary device to collect GPS and telemetric events in a vehicle such as speed, ignition on/off, broom up/down (street sweepers), compactor activation (trash trucks), and pump activation (gasoline-dispensing trucks). An ArcIMS viewer relays the data to monitors and managers, representing various departments and functions throughout the enterprise. Using an

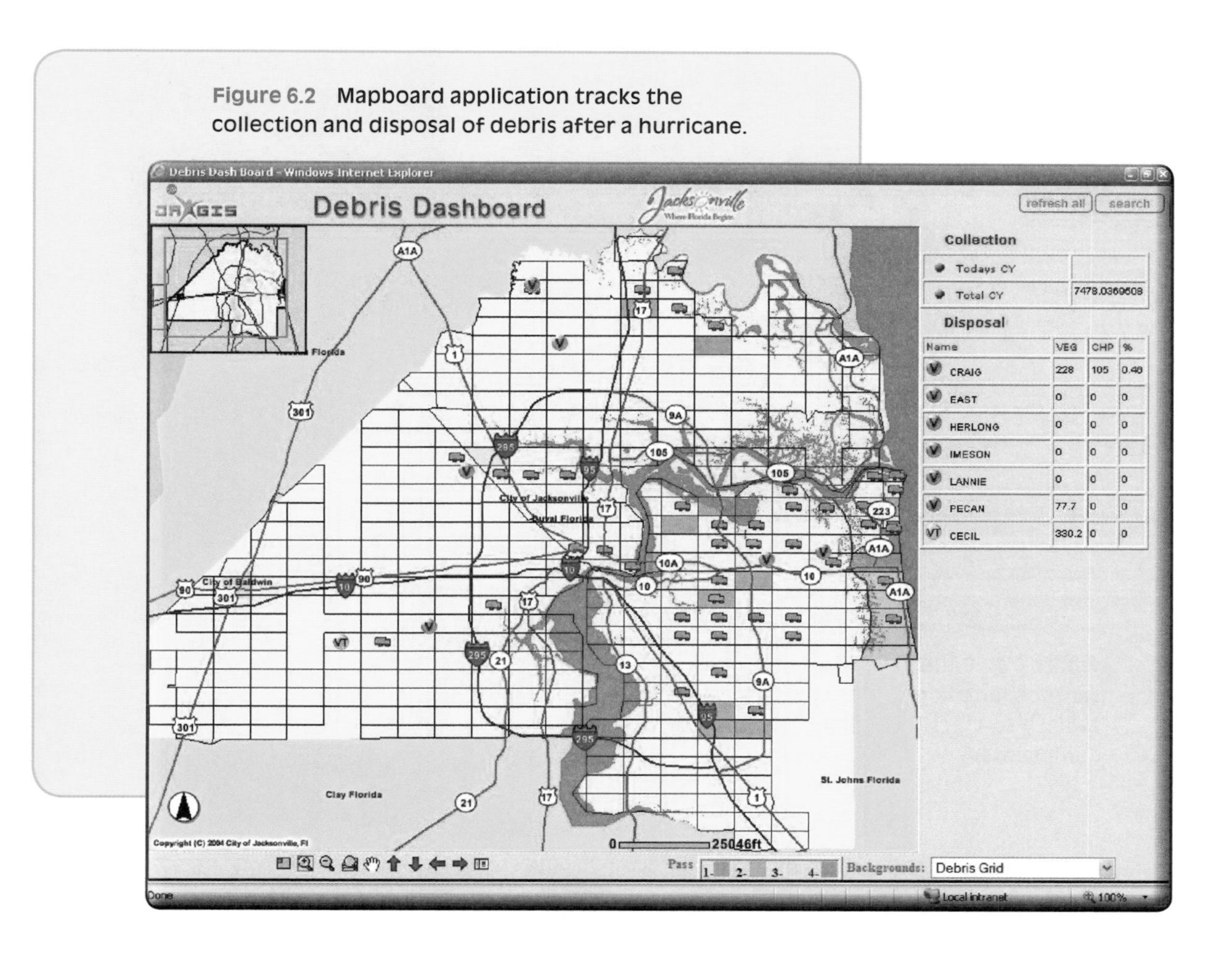

Figure 6.2 Mapboard application tracks the collection and disposal of debris after a hurricane.

automatic polling rate of one poll per twenty seconds per vehicle, the system has processed more than three million records since March 2006.

Collecting mobile data has enabled administrators and fleet managers to evaluate and analyze information relevant to the operation. Results are having a positive direct impact on decisions regarding risk management, resource allocation and distribution, asset maintenance, and route optimization.

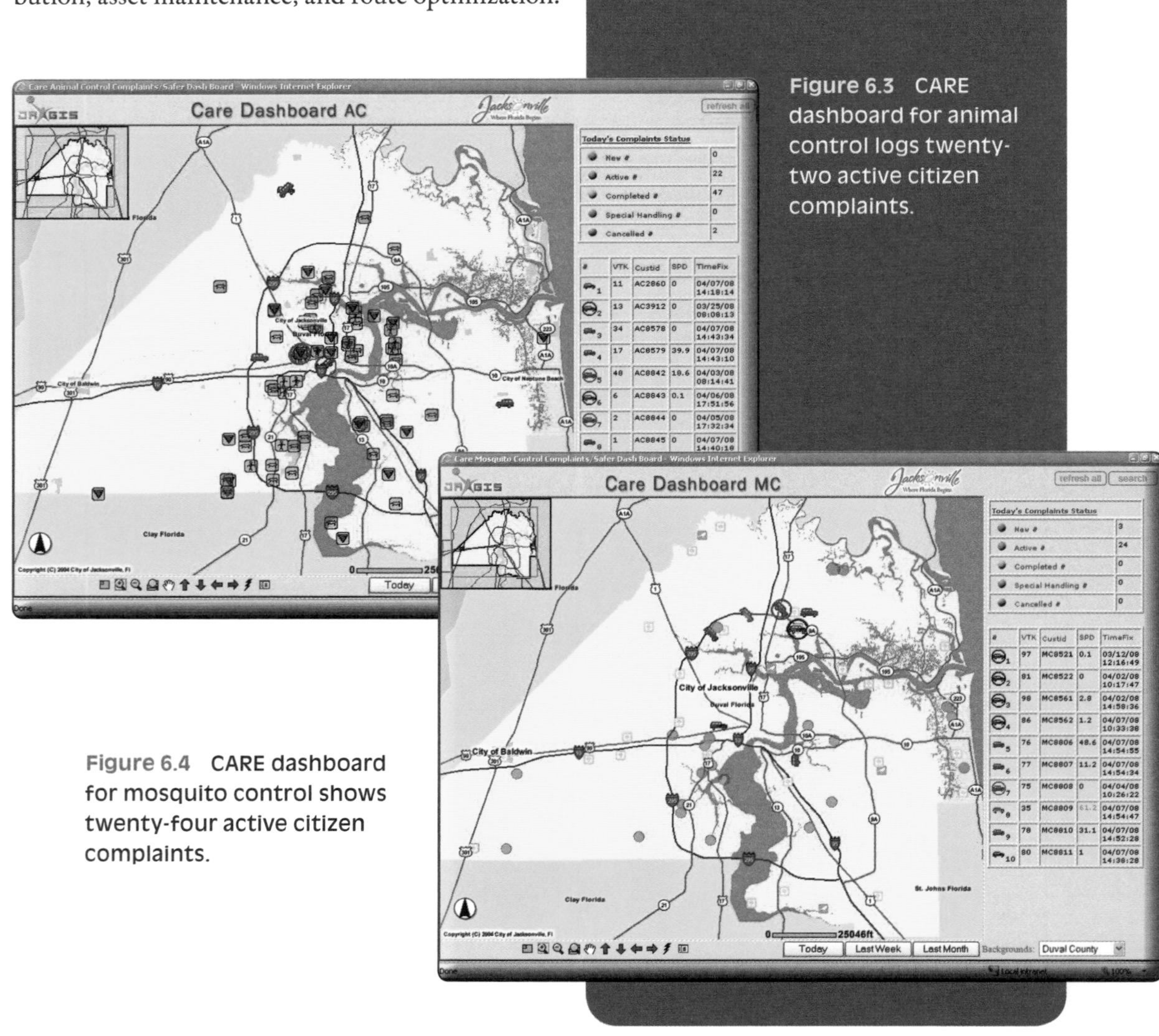

Figure 6.3 CARE dashboard for animal control logs twenty-two active citizen complaints.

Figure 6.4 CARE dashboard for mosquito control shows twenty-four active citizen complaints.

County in Iowa finds multiple uses for digital data

CHAPTER: Making decisions on the fly
ORGANIZATION: Pottawattamie County
LOCATION: Council Bluffs, Iowa
CONTACT: David Bayer, GIS coordinator
david.bayer@pottcounty.com
PROJECT: Integrating assessor's data for emergency responders
SOFTWARE: ArcGIS and ArcSDE
ROI: Cost and time savings; increased efficiency, accuracy, and productivity

By David Bayer

When emergency calls come in, responders have to make decisions on the fly, and the choices they make in deploying personnel, reaching an incident, or making a plan of attack are crucial in saving lives. While their tasks are demanding, GIS technology is making inroads in helping to ensure their success. Pottawattamie County, Iowa, has implemented a helpful solution for its emergency responders at no added cost.

Pottawattamie County is directly east of the Missouri River and Omaha, Nebraska. The county encompasses nearly 1,000 square miles with a population of approximately 90,000. In 1998, the county government made a significant investment in GIS technology, including aerial photography, various spatial datasets, powerful new workstations, and software. By 2002, the Board of Supervisors recognized the potential of GIS and subsequently addressed the need for internal management and budgeting. An independent county department was formed, and capital and budgeted monies were set aside to fund the new department. In June 2002, a GIS program manager was hired, and the GIS department was officially formed to administer and grow the county's GIS program.

The county has a 911 Communications Center that plays a critical role in the delivery of all police, fire, and medical services to the county. It implements a core off-the-shelf dispatching tool called GeoLynx. This software uses the county GIS department's base data, including aerial photography, centerline, municipal boundaries, political townships, bike trails, emergency-medical-services districts, fire districts, and law districts. When a landline or wireless call comes into the center, the map zooms in on the appropriate location using the centerline feature class and its address-range

Figure 6.5 The government seal of Pottawattamie County, Iowa

data. With this information, the appropriate safety service can be dispatched to the call location.

GeoLynx is dispatching software developed by GeoComm. It uses ESRI's MapObjects software and provides fully automated E911 emergency-dispatch mapping information for both wired and wireless calls. The county's GIS data is housed in an enterprise geodatabase using Microsoft SQL Server and ArcSDE and is maintained using ESRI's ArcGIS Desktop software.

The dispatch system has the ability to attach images to an address. The 911 Communications Center manager, Chris Moore, recognized the value of using photos and diagrams for display in the county address database when a call comes into the center. The process of capturing digital photographs and diagrams of all buildings in the county was already in place in the Assessor's Office, so incorporating the images only had two requirements: rename the photo and diagram files to their associated parcel address and then load them into the ImageLynx database.

Renaming the files with their associated address required mirroring address standards for the two address databases. The two address databases included street-centerline data and parcel-address data. GeoLynx uses the centerline addresses to locate calls, but the photos and diagrams are linked to parcels by their parcel identification number (PIN). These two databases had differing standards. For example, the street-type fields had different abbreviations. An avenue in one database might be AVE while it was AV in the other. If a photo was renamed to 515 5TH AVE.jpg (the parcel address for that photo), but the call came in as 515 5 AV (the centerline address), the software would not

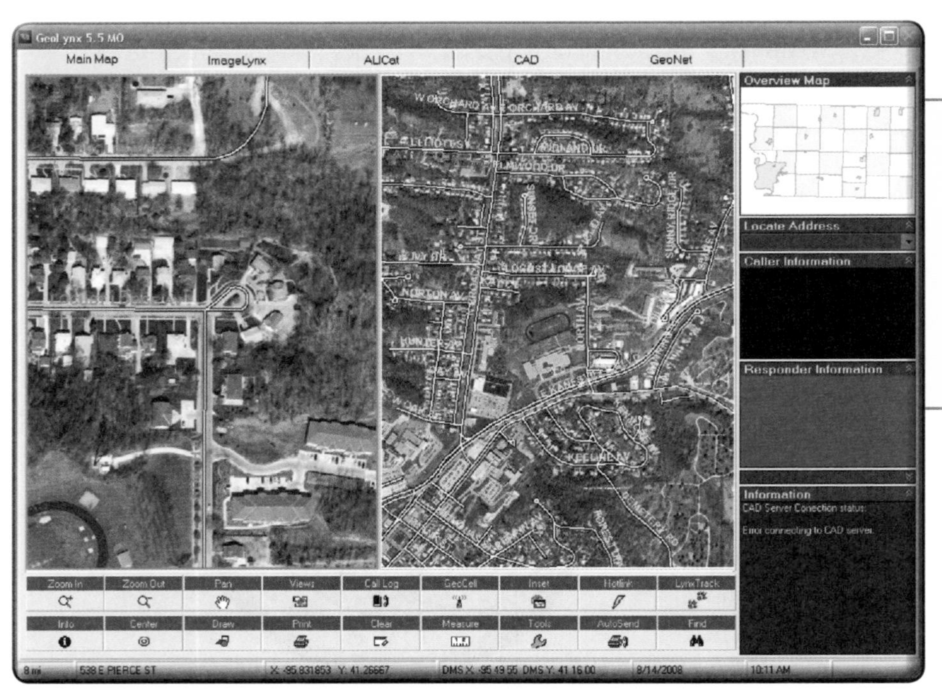

Figure 6.6 GIS software is used to field all police, fire, and medical-service calls that come into Pottawattamie County's 911 Communications Center dispatch.

recognize that there was a photo of the house available for that location. To resolve this problem, one standard was created, and the two databases were revised to match those standards.

The second requirement was to build the database consisting of two tables. The first table contains all of the addresses that have at least one associated photo. The second table contains all of the photos and diagrams and their names. The two tables have a one-to-many relationship that enables each address to have multiple files. A script was written to capture all of the file names in a directory and load those names into the database. Finally, multiple tests were performed on a development server before the project was implemented in the production environment.

This project was completed at no additional cost. The software was already in place and in production, and the capture of the digital photos and diagrams was already a proven process. Plans call for incorporating floor plans of buildings where large numbers of people gather, such as schools, hospitals, the convention center, and casinos. Eventually, photos will be sent directly to the first responder to be viewed in the police car or the fire truck.

An added benefit to the project was the resulting collaboration among four county departments. The Assessor's Office regularly captures the photos and diagrams. The GIS department maintains the street centerline and parcel addresses and the other base layers that the 911 Communications Center uses. The Information Technology department performed the bulk of the workload by renaming the images and building the database. The 911 Communications Center dispatcher is the end user of the product. Having the photo of the building with a quick diagram of its perimeter enables the dispatcher to better describe the destination to public-safety personnel, which saves valuable seconds in an emergency.

Figure 6.7 Homes in Pottawattamie County are kept safer with GIS software that provides emergency workers with a digital photo of buildings and a diagram of their perimeter.

Having the photo of the building with a quick diagram of its perimeter enables the dispatcher to better describe the destination to public-safety personnel, which saves valuable seconds in an emergency.

GIS streamlines sewerage management

CHAPTER: Making decisions on the fly
ORGANIZATION: Zanesville/Muskingum County Health Department
LOCATION: Zanesville, Ohio
CONTACT: Bob Brems, epidemiologist
bbrems@zmchd.org
Ed Shaffer, supervising sanitarian
eds@smchd.org
PROJECT: Incorporating GIS in sewerage management
SOFTWARE: ArcGIS, ArcView
ROI: Cost and time savings; improved access to information

By Bob Brems and Ed Shaffer

As the population growth in southeast Ohio continues a steady movement from cities to rural areas, the extension of public sewers is not keeping pace. In many cases, extending public sewers to remote or sparsely populated areas is not economically feasible. As a result, the individual, home-based sewage-treatment system has taken a prominent place in the overall practice of sewage treatment, and is used extensively throughout Muskingum County, Ohio.

The first step in the design of a home sewage-treatment system is to determine the suitability of soils. Most home treatment systems depend on soils to both treat and absorb wastewater. Factors such as soil permeability; depth of seasonal groundwater, bedrock, or other limiting layers; surface topography; and the flow of runoff water all must be evaluated in determining the suitability for an on-site sewage treatment system.

The most common, and economical, type of system consists of a septic tank with a series of leaching lines. The tank serves as the primary treatment. It allows for the settling and storage of most of the solids and starts the anaerobic breakdown of the wastewater. The leaching lines disperse the wastewater over a large area allowing further treatment and absorption by the soil. Other types of systems have been developed to overcome limitations of the soil or to operate on smaller home sites. These systems include aeration; filtering water through sand, peat or other material; and chemical disinfection.

Muskingum County, Ohio, has more than 10,000 existing systems of record, and more than 300 new systems are installed each year. To better manage alterations to existing systems and new installations, a GIS was implemented that enables Zanesville/Muskingum County Health Department (ZMCHD) sanitarians to quickly assess a property's suitability for a sewage-treatment system.

Methodology

ZMCHD does not generate and maintain base geographic files of the county and has worked closely with the Muskingum County GIS department to obtain current geographic files. An accurate centerline file with address ranges provides the most use, because most health department datasets contain an address. In the past, ZMCHD relied on TIGER files for most GIS projects. Muskingum County was one of the first counties in Ohio to

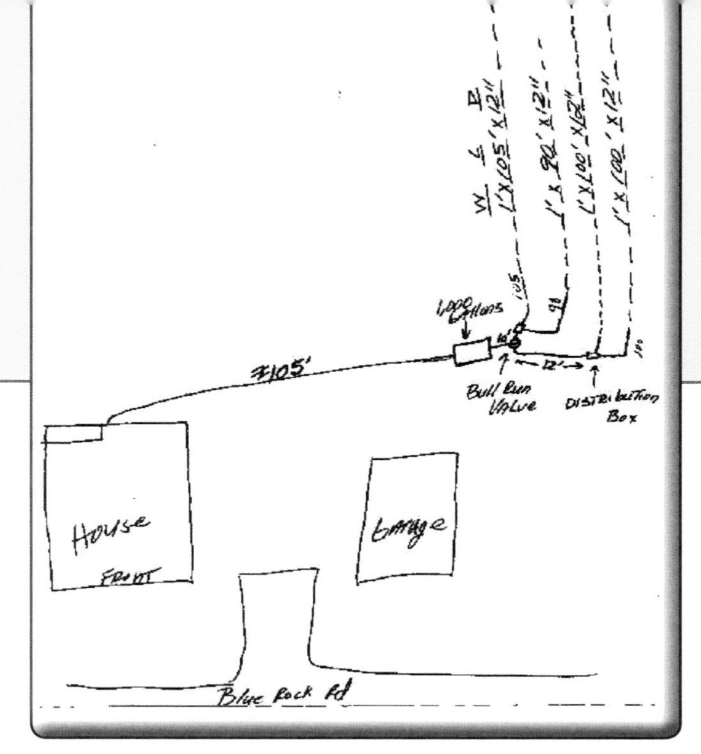

Figure 6.8 A scanned sketch of a homeowner's sewage system can be accessed using a map hyperlink.

take advantage of the Location Based Response System (LBRS) initiative, a state effort to share the county's cost (determined by the road miles and addressable structures in the county) of obtaining accurate, field-verified road-centerline data with address ranges. Additionally, every residence's road access (i.e., driveway) was also collected, so the data includes a field-verified address point for every residence in the county. The health department is able to accurately geocode its data using these point and centerline files. Other local geographic data such as land parcels, contours, spot elevations, orthophotography, floodplains, and soil composition was used to implement the sewage-treatment-system GIS.

Using a Microsoft Access dataset of sewage-treatment-system information, existing system locations were geocoded using ArcGIS. The goal was to locate a point as close as possible to the actual sewage-treatment system using existing data. To accomplish this, two geocoding services (address locators in ArcGIS) were created. The first geocoding service was created using the Single Field (file) style. The reference data used was the LBRS address point file, and the Key Field was the field containing the address consisting of house number and street name, labeled LSN in the dataset. System defaults for matching options of 80 for spelling sensitivity, 10 for minimum candidate score, and 60 for minimum match score were used. Output fields of x,y coordinates and Standardized Address were also selected. The geocoding service was run matching the sewage-treatment-system address, labeled Address in the dataset, and saved as a point shapefile.

The second geocoding service was created using the U.S. Streets with Zone (file) style with the LBRS centerline file as reference data. Fields were matched to the LBRS centerline file for House from Left, House to Left, House from Right, House to Right, Street Name, Street Type, Left Zone, and Right Zone. System defaults for matching and output fields were the same as in the first geocoding service, so that the two resulting shapefiles could be easily merged.

The first geocoding service was run enabling a point to be placed on the property where the sewage-treatment system is actually located. It is understood that it is not the actual location of the system on the property, only that it is matched to the property. The systems that could not be matched to the address point were then matched using the second geocoding service to approximate the system's location on the road centerline. Unmatched records were reviewed to determine why geocoding by either method was not successful.

A scanned sketch of the system layout, as documented by the licensing sanitarian, is included in the sewage-treatment-system database. The ability to locate the sewage-treatment system with a point allowed ArcView's hyperlink feature to be used. The sanitarian reviewing the sewage-treatment system can view the sketch without having to retrieve the paper record. The scanned image was included in the Access dataset as an embedded OLE object. Extracting the path location of the sketch required the use of a Visual Basic program so the hyperlink feature could be used.

Results and considerations

Of the 10,426 existing sewage-treatment-system records, 8,386 (80.4 percent) were successfully geocoded to the address point file. Manual geocode matching was performed to place as many

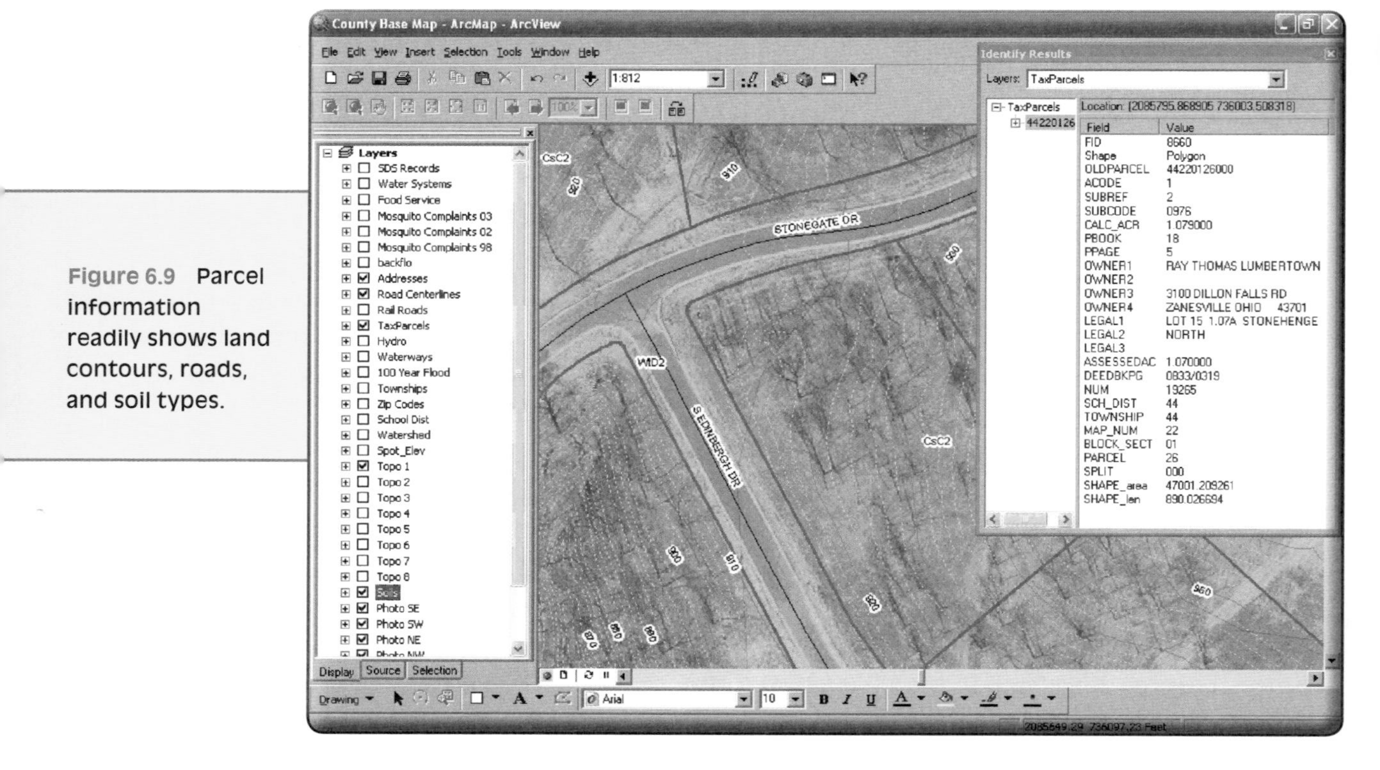

Figure 6.9 Parcel information readily shows land contours, roads, and soil types.

The GIS enables sanitarians to perform sewage-treatment-system reviews of existing systems in minutes. These reviews are necessary if there is a complaint or if a homeowner wishes to modify the system or request a variance.

systems as possible at the correct address point. This was done to compensate for typographical inconsistencies between the LBRS centerline file and address information in the health department's sewage-treatment-system database. The remaining 2,040 records were geocoded to the address-verified centerline file to approximate the system location. A total of 1,801 (17.3 percent) systems were successfully geocoded to the centerline file, leaving only 239 (2.3 percent) that could not be geocoded. Upon review of the unmatched records, most were found to be older records that lacked an address or were only designated by a lot number or street name.

While detailed time-saving data is not available, the sewage-treatment-system GIS has reduced the number of man-hours needed to manage the program. Using the GIS has eliminated the need for the sanitarian to review printed parcel maps only available in the Auditor's Office located in another building. It has reduced the time needed to leaf through paper files to review system configuration. Some site visits have even been eliminated because of the ability to assess property information using GIS.

Improving access to data (e.g., system sketches and land-parcel information) reduces the amount of time that environmental-health staff must spend reviewing case records. The GIS enables sanitarians to perform sewage-treatment-system reviews of existing systems in minutes. These reviews are necessary if there is a complaint or if a homeowner wishes to modify the system or request a variance. Some homeowners request a variance for special circumstances, which must be approved by the health board. The ability of the health board to make decisions regarding variance requests is greatly enhanced with a map easily generated using GIS.

GIS is especially useful as new sewage-treatment systems are installed. Since current land-parcel information is available, initial evaluations as to the size of the lot and location relative to floodplains can be made quickly. With soil composition, orthophotography, land-contour, and spot-elevation information, initial assessments about system location can also be made. By having GIS as a visual tool, sanitarians can now have detailed phone consultations with property owners and contractors regarding the placement of homes and sewage-treatment systems. These preliminary conversations ordinarily would be difficult without visiting the site first. In fact, many of the property owners do not reside in the county and are building retirement or weekend homes, so being able to discuss issues with them without having to arrange a meeting at the site saves time and improves customer service. Additionally, GIS will allow the sanitarian to quickly reference other systems in proximity, which can aid in the design of the system.

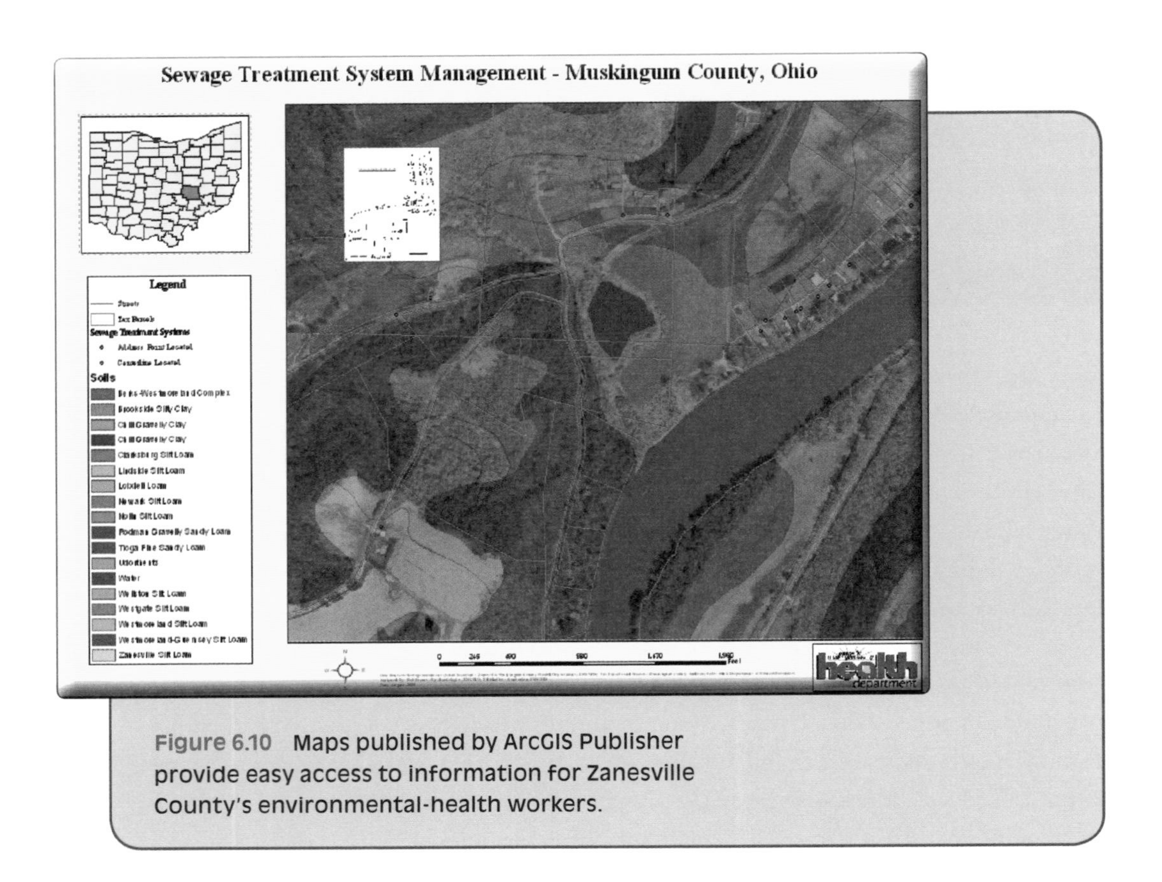

Figure 6.10 Maps published by ArcGIS Publisher provide easy access to information for Zanesville County's environmental-health workers.

GIS is a valuable tool for sewerage management. With it, sanitarians can quickly use geographic information critical to decision making, eliminating the need to refer to cumbersome printed maps. While it does not eliminate all field visits, it does make field visits more productive, because the sanitarian can become familiar with the property (size, contours, soil) beforehand.

Making the map accessible using ArcGIS Publisher exposes more staff to GIS. As staff members become more familiar with the technology, it is anticipated that future GIS projects will be undertaken. Ideas will be generated about which programs might benefit from GIS analysis, and since most public-health data contains an address, geographic references are available.

Challenges remain. This project implemented an existing database of sewage-treatment-system information that is maintained separately from the GIS. Therefore, new sewage-treatment-system installations are not incorporated into the GIS unless a separate geocoding project is undertaken. Integrating the sewage-treatment-system database into the GIS using a geodatabase is a logical future step. Efforts to keep the same workflows for staff responsible for inputting sewage-treatment-system data is a priority. This must be accounted for in the integration process, and ideally, an automatic geocoding process could be incorporated so new systems are immediately available in the GIS after they are entered into the database. This will require some outside expertise to fully implement.

Acknowledgments

Thanks go to the Muskingum County auditor and GIS department for having the vision to invest in the LBRS project. Without accurate road and address information, this project (and future ones) would be difficult.

GIS is especially useful as new sewage-treatment systems are installed. Since current land-parcel information is available, initial evaluations as to the size of the lot and location relative to floodplains can be made quickly.

Supporting policies with GIS

7

Built on a foundation of ideals and values, public policy serves as the basis upon which governments make decisions, take action, or choose inaction. The need to formulate policy can stem from a number of reasons or sources. Sometimes policy processes are initiated in response to issues that emerge from a crisis or emergency, as a consequence of another government's decision, or as a way to allocate resources. Policy can also be generated to meet stakeholders' or public concerns or as a reaction to untoward attention. Some issues are brand new, and the problems to be addressed have not been clearly articulated or documented, while others are known but lack viable solutions. In some cases, policies are in place, but their implementation is problematic.

Public-policy development generally wends its way through defining the problem, describing goals, fleshing out alternative scenarios, agreeing on a policy, and implementing and evaluating it. This process can be contentious. Outcomes of policy development can benefit some and become a painful experience for others. Successful policy implementation and support requires rational strategies that are based on relevant, persuasive data.

Having the capability to use GIS technology and spatially referenced data has become an essential and invaluable asset in supporting public-policy and strategic-planning decisions and enhancing government management and operations. When a group or organization uses a common language such as geography to exchange ideas, that group develops a common bond and begins to develop a strong basis for communication and collaboration. Defining GIS technology as a policy tool by which outcomes will be pursued, information exchanged, regulations monitored, and expenses calculated will help officials deliver the best outcomes while increasing productivity, providing better services, enhancing smart-growth decisions, and empowering the community.

The availability of robust, accurate GIS data and the capability to analyze the data makes easier work of outlining and determining how the policy will be carried out. Leaders and decision makers in every discipline, including public-health analysis, disaster preparedness, sustainability, economic analysis, and budget and finance analysis, find that GIS software is a valuable visualization tool that can be used in conjunction with ancillary information to develop and understand policy goals and implementation and to target areas for enforcing that policy.

GIS helps pick up the pace in laid-back Curry County

CHAPTER: Supporting policies with GIS
ORGANIZATION: Curry County GIS Department
LOCATION: Curry County, Oregon
CONTACT: Toni Fisher, GIS coordinator
fishert@co.curry.or.us
PROJECT: Curry County GIS
SOFTWARE: ArcIMS, ArcView, ArcSDE
ROI: Automated workflows

By Toni Fisher

Situated in a beautiful location along the Pacific Coast in the southwest corner of Oregon, Curry County has a reputation for its relaxed pace. Initially settled along the coast, Curry County relied on commerce that was dependent on water transport in its early days, and slow development of inland transportation routes has contributed to the county's rural atmosphere and sparse population.

Curry County's laid-back style also carried over to its business practices—so much so that many of the procedures in county offices had not been updated in twenty-five years. The gap between its business practices and standard computerized methods was widening as a result. Unable to attract highly skilled employees because of a lack of resources to pay competitive salaries, the county fell behind in administrative efficiency over the years. The county did not make use of computer technology to help with daily tasks, and paper-based records and departments that operated independently of each other was the norm in county offices.

Curry County has reason to be proud of the significant transformation it has undergone recently because of GIS technology. Paper-based records are difficult to search, time consuming to file, available to only one user at a time, and stored in a single location. Since its adoption of GIS in 2006, the county has been converting its paper-based system to digital data and has realized considerable benefits in time-saving and data-sharing abilities.

Forming partnerships for the common good

Accurate digital basemaps of the county had not been developed until the GIS initiative began, which was possible with funding through partnerships that provided for the production of a new countywide orthophoto and digital terrain model. From these, road and hydrology layers were derived, enabling the first major GIS project to commence. Data is expensive to produce, maintain, and manage, and the Curry County partnerships have enabled shared funding of one common database used by all. ArcIMS and ArcSDE have provided many users access to the common dataset. Without sharing, no one agency within county limits could have afforded by itself to produce, maintain, and deploy the data.

Using the recently created basemap data, in conjunction with the Bureau of Land Management,

U.S. Forest Service, Oregon Department of Forestry, and Curry County Emergency Services, the county's GIS coordinator set up a wildfire evaluation project to facilitate data collection for homes potentially at risk. Collecting data in the field and inputting it directly into the GIS is more efficient and accurate than collecting the same information on paper. Digital data lends itself to decision-making support.

Figure 7.1 A public workshop keeps Port Orford residents informed through GIS of changes to the community-refinement plan.

In 2002, a major wildfire burned nearly 500,000 acres and several homes within the county. Protective services during the fire were greatly hampered by the lack of available information. After the fire, an effort was made to gather data, but without GIS storage capabilities and standard collection methods, the paper-based collection of data proved to be of little help. With the addition of GIS resources, Oregon Department of Forestry workers, equipped with laptops and ArcView software, collected data during summer 2005 for 1,000 residences within the county. Using geodatabase domains and subtypes made data entry easy for personnel with little or no computer literacy. The project continued through summer 2006 and was scheduled to be completed that fall. At the end of data collection, the goal was to prioritize areas needing attention to ensure public safety in the event of wildfires.

Curry County's collaboration with the wildfire preparation team is only one of the partnerships forged to fund GIS and share a common GIS database across a largely rural county. The City of Port Orford, with a population of 1,153, is one of three cities in the county. Poised to grow, this coastal city needed to review its comprehensive "refinement plan" and determine zoning changes in conjunction with public input. The Oregon Department of Land Conservation and Development had set aside funding assistance for Port Orford's efforts. Instead of paying a consultant to develop mapping for the project, the city joined the county GIS partnership. The county GIS has provided city officials with answers to their spatial and database queries, hard-copy maps, and GIS demonstrations at public meetings. GIS played an integral part in the decision-making process for the refinement plan released in fall 2006.

Figure 7.2 A GIS worker collects data for home evaluation in Curry County.

Public education in GIS via public demonstrations contributes to the development of mainstream acceptance and literacy in GIS, directly affecting the future ability of local governments to budget for enhanced GIS technology. Evaluation of Port Orford's refinement plan enabled this kind of education.

Curry County GIS has an ArcIMS internal Web site accessible to all county offices and Curry County partners. With easy, affordable access to the data, many departments are updating their business practices and making the change from paper and rows of filing cabinets to digital, online data. Curry County has used GIS to jump-start progress, promote partnerships, integrate departments, and improve public services.

Collecting data in the field and inputting it directly into the GIS is more efficient and accurate than collecting the same information on paper. Digital data lends itself to decision-making support.

GIS for healthy Georgia communities

CHAPTER: Supporting policies with GIS
ORGANIZATION: Georgia Department of Human Resources, Division of Public Health
LOCATION: Atlanta, Georgia
CONTACT: Elaine Hallisey, director of GIS team ehallisey@dhr.state.ga.us
PROJECT: Making community health assessments
SOFTWARE: ArcGIS, ArcMap, ArcReader
ROI: Cost and time savings; increased efficiency, accuracy, and productivity; enhanced communication and collaboration; more efficient allocation of resources; improved access to information

By Elaine J. Hallisey

Within the Georgia Department of Human Resources, the Division of Public Health (DPH) is the lead state agency ultimately responsible for the health of the population. DPH recognizes three core functions of public health: monitoring health in communities (assessment), developing public policy, and ensuring access to and evaluation of care (assurance).

The Office of Health Information and Policy (OHIP) leads the health-assessment component within the division. OHIP's purpose is to provide valid and reliable evidence about the health status of the population of Georgia. The evidence of health status and other health-related phenomena provided by OHIP is used to direct the division's policies and actions.

In its role of health assessment, OHIP provides data and information regarding the relationships between various measures of health and the socio-economic status of a community. The goal is to determine if there are patterns that would be helpful in developing policies that target programs and allocate resources.

Measures of health

DPH chose to explore measures of health that represent the seven leading causes of years of potential life lost (YPLL) due to death before the age of 75. YPLL was chosen as a realm of study, because health issues affecting people under 75 produce a greater economic and social burden on society. In addition, these health problems can be managed by DPH and decreased through special programs and education. The leading causes of YPLL in Georgia are, in this order:

- Motor-vehicle accidents
- Lung cancer
- Heart disease
- HIV/AIDS
- Suicide
- Homicide
- Stroke

> **The goal is to determine if there are patterns that would be helpful in developing policies that target programs and allocate resources.**

Data for these measures is gathered via death records that are maintained in a standardized repository. GIS analysts examined additional measures of health of special interest, including substantiated infant maltreatment, teen births, and infant mortality. All mortality and other health records are routinely geocoded to the individual's residence address so that mapping of each measure of health is a straightforward process.

Demographic profiles

The demographic profiles of Georgia are the result of a segmentation analysis similar to those used by many marketing and polling agencies. To reduce costs and to maintain data control and quality, DPH staff decided to create custom profiles from numerous socioeconomic variables rather than purchase existing profiles from a commercial firm.

Demographic profiling is a means of reducing data into distinct categories with unique properties so that many rows of individual records are accumulated into a few categories with specific properties. Relevant measures of health can be analyzed in terms of the categories, and the determination of variation in the measures of health by category can be discovered.

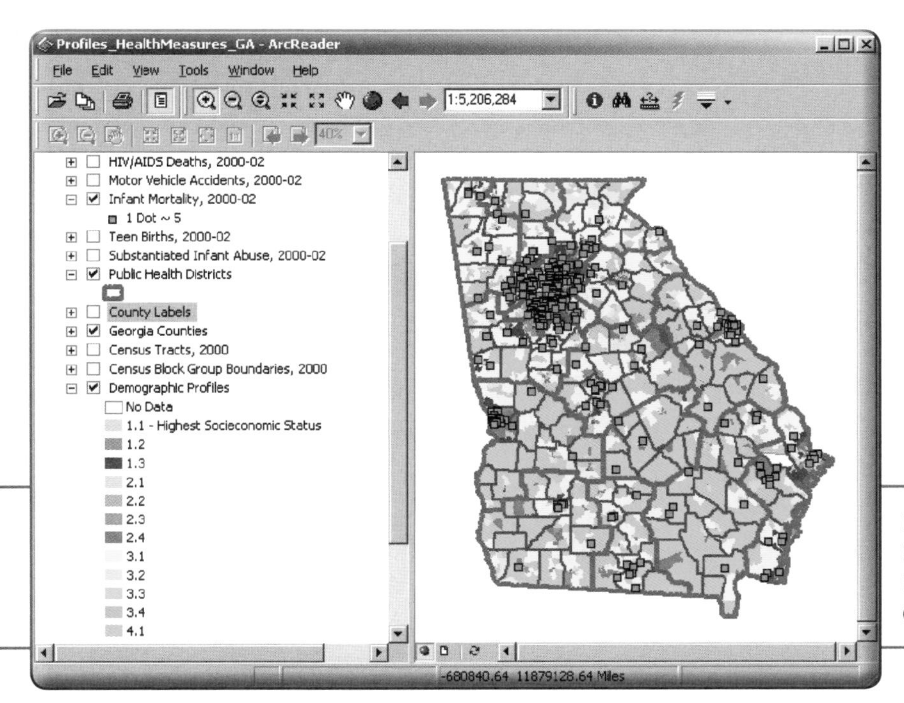

Figure 7.3 ArcReader interface shows the infant mortality rate in relation to demographic profiles.

The profiles were developed in a statistical software package from a variable set of socioeconomic data at the census-block group level. The set consisted of the following:

- Age distribution as a percentage of the population
- Education as a percentage of the population age 25 and older
- Occupation as a percentage of the population age 16 and older
- Family structure as a percentage of households
- Number of dependents
- Population density
- Housing features (vacancy, rented units versus owned units, length of residency, age of housing, and housing value)
- Median household income
- Median number of vehicles per household
- Urban versus rural population

Four major socioeconomic clusters were initially defined: prosperous, middle class, lower middle class, and lower income. From these four major clusters, eighteen demographic clusters were developed. Using ArcMap software, a GIS analyst joined the assigned demographic profile to each block group, creating a new shapefile with a standard legend. Detailed descriptions of each of the eighteen profiles, from the prosperous blue areas through the middle-class greens and yellows to the lower-income reds and purples, are available on the DPH Web site (http://health.state.ga.us/demographicprofiles/index.htm).

All mortality and other health records are routinely geocoded to the individual's residence address so that mapping of each measure of health is a straightforward process.

Georgia's Eighteen Demographic Clusters	
Demographic clusters were created from census data variable classes containing age, income, family structure, housing value and housing type, education, and employment type. These variables were analyzed using a classification model that created four major groups, which further partitioned into a total of eighteen distinct demographic clusters. The derived socioeconomic status is arranged from "higher" to "lower" within the four major groups and their respective demographic clusters.	
1.1	Georgia's wealthiest cluster is primarily populated by "new money" executives and professionals living in tract mansions of metropolitan suburbs and exurbs. Predominantly white with a high index for Asians, this highly educated cluster is composed of married couples in their 40s and early 50s with adolescent children.
1.2	This well-educated, suburban cluster, dominated by professionals and managers, has the second-highest level of affluence in the state. Mostly white with a high index for Asians, they are older than cluster 1.1 and include married couples with adolescent and grown children.
1.3	Found in the metro suburbs, this mixed-ethnicity, more youthful cluster is populated by married couples in their 30s and early 40s with young children. The majority have some college or are college graduates. Most are employed in managerial and other white-collar jobs, while some are high-earning blue-collar families. This cluster enjoys a median income well above the state average.
2.1	This cluster is characterized by its high concentration of highly educated young people in their late 20s and 30s renting in upscale urban neighborhoods. Dominated by white and Asian married couples without children, this cluster is positioned to join prosperous families in the next decade.
2.2	This cluster is primarily populated by people in their late 30s and early 40s and older people over 65. A mixed-ethnicity group, they live in rented apartments and condos in urban areas; and although many are college educated, their median incomes are well below cluster 1.3 and 2.1. Children are not highly represented in this cluster.
2.3	This is a very young cluster of mixed ethnicity living in middle-range value apartments in urban/suburban areas. Many are college educated or have some college, and their income is exactly the state average. They have an elevated index for single-parent families with children, although the population under 18 is small.
2.4	This mixed-ethnicity cluster represents the college, military, and prison populations in Georgia (those populations living in group quarters). They are mostly under 24 and have lower incomes than the state average.

Table 7.1 State of Georgia Demographic profiles. *(continued on next page)*

Georgia's Eighteen Demographic Clusters (continued)

Cluster	Description
3.1	This cluster is a white, middle-class rural cluster dominated by married families with children. They are mainly home owners, but the value of their housing is much lower than in urban areas. Many in this cluster are high school graduates, but they are higher than the state average in population that failed to graduate from high school. This cluster is highly represented in farming and construction and is widespread across the state.
3.2	Although this cluster includes younger populations, it is dominated by the 55+ age group. Found predominantly in N/NE rural counties of Georgia, the cluster is white with some African-American population. As would be expected of a population with many persons living on fixed incomes, this cluster has lower income than families currently working, but their incomes are still average compared to the state.
3.3	This mixed-ethnicity cluster is average in the age profile but has a higher percentage of single-parent families than the state as a whole. A large percentage did not finish high school, and they are much less likely to have a four-year college degree. Approaching the state average in income, these families work in lower-paying service, sales, and managerial jobs to maintain a lower-middle-class lifestyle.
3.4	Composed of rural married and single-parent families, this cluster is older than cluster 3.1 and less affluent. Predominantly white with some African-American population, this group is much more likely to own low-value housing, not to have finished high school, and to work in farming.
4.1	This cluster is composed of newly arrived immigrants to the United States. Primarily Hispanic, the cluster is young and not well educated. Dominated by single households, but with a substantial percentage of married families with children, this urban population lives in rental housing, is below average in income, and works in service, construction, and processing industries.
4.2	An urban cluster, this African-American group has a high representation of elderly people and single-parent families with children. Not well educated and with lower-than-average incomes, this group lives in areas with high vacant housing and low housing values. Although poor, this cluster also demonstrates social stability with almost 60% showing home ownership and 30% being married family households.
4.3	This is a young cluster with a high proportion under 24. Primarily African-American and with a high index for Hispanics, this cluster is characterized by singles and single-parent families with children living in urban/suburban rental units. They work in service jobs, and their income is more than 30% below the state average.

Georgia's Eighteen Demographic Clusters (continued)	
4.4	Found in old mill towns in suburban and rural areas, this cluster is composed predominantly of African-American married families and single parents with children. The population is skewed to the very old and very young. They are primarily renters, have high school or less than high school educations, and work in service industries, making half the state average in income.
4.5	This African-American cluster is much like cluster 4.4 but is more urban, older, less educated, and lower in income. They are more likely than 4.4 to live in rental units.
4.6	This is a very small and unusual urban cluster. It is dominated by an African-American population with a high percentage of whites. The cluster is more than twice the state average in population over 65 and has very few children. Oddly, this cluster has more males than females for the ages 18–54. The group lives in rental units, is very poorly educated, and experiences very low income.
4.7	This is a very young African-American cluster composed of single-parent families with children. The population under 18 is very high, and there is almost no elderly population. The cluster is poorly educated, lives in rental units, and has the lowest median income in the state.

Visualization in a published map file

A GIS analyst created a map document in ArcMap displaying the demographic profiles as a base layer and each of the measures of health as point layers. Within the DPH framework, individual points are shown, but for public viewing, the measures of health were symbolized in dot-density form, with each dot representing five events, to preserve confidentiality (http://health.state.ga.us/programs/ohip/arcreader.asp). Using ArcGIS Publisher software, the map documents were converted to ArcReader format. With ArcReader, DPH can freely distribute interactive interfaces that enable users to design, print, and copy and paste their own maps. The user is able to visually analyze the data, including zooming in and out, turning map layers on and off, and performing simple queries and distance measurements.

Users can examine how the various measures of health and demographic profiles are related spatially. They may note that any correlation present among the measures of health and demographic profiles seems to be more readily visible in the urban areas of the state, as opposed to rural areas. This could be because of the large differences in population density between the areas.

Focusing on the metropolitan Atlanta area, where population density is uniform, HIV/AIDS deaths are highest in the lower-socioeconomic areas. Homicide and suicide present an interesting juxtaposition. Homicides fall primarily within the lower socioeconomic areas, whereas suicides occur in middle-class and prosperous areas. Substantiated infant abuse displays a pattern similar to HIV and homicide, occurring mostly in the lower-socioeconomic neighborhoods.

Figure 7.4 Blue squares representing HIV/AIDS deaths are more prevalent in the lower-socioeconomic (red) areas of Atlanta.

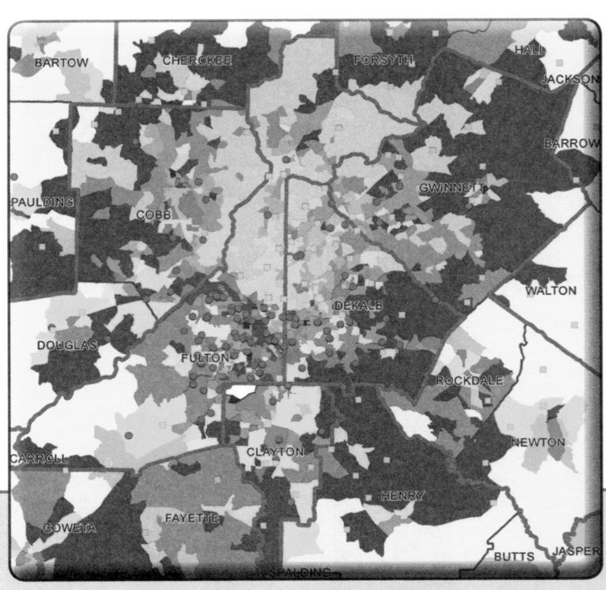

Figure 7.5 Homicides (green circles) occur in lower-socioeconomic areas of Atlanta, whereas suicides (orange squares) tend to occur within middle-class and prosperous (blue) areas.

Northern Fulton County, which has a low rate of substantiated maltreatment, is in stark contrast to south Fulton, which has a high rate of substantiated maltreatment. Northern DeKalb County, immediately adjacent to Fulton, also has a lower rate than southern DeKalb. Recognizing the visual patterns apparent in infant maltreatment data led to further statistical analysis beyond the visualization arena. Statistical analysis demonstrates that there is an exponential correlation between the rate of infant maltreatment and the socioeconomic level of the eighteen demographic profiles.

Policy development

Policy implications from population segmentation are far reaching, with two primary dimensions. One focuses on the discovery of fragile populations whose rate of poor education (high school or less within populations age 25 and older) is greater than 70 percent (Millard 2005). A high burden of poor education tends to sustain lower socioeconomic status and dependency on social services. The lower-socioeconomic groups in demographic profiles 3.4 through 4.7, primarily, exhibit this property. A striking difference between segments 3.4 and the lower 4.1 through 4.7 is population density, making those segments in urban areas extremely

Recognizing the visual patterns apparent in infant maltreatment data led to further statistical analysis beyond the visualization arena.

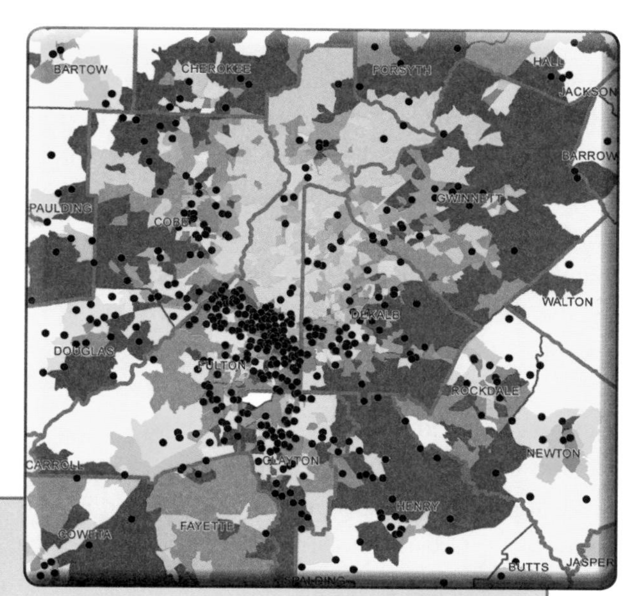

Figure 7.6 Cases of substantiated infant abuse (black circles) fall primarily within lower-socioeconomic (red) areas.

fragile. Their low population combined with high density may concentrate disease (e.g., HIV) and dysfunctional behaviors (e.g., homicide).

The second dimension relates to the strategic marketing of health information to members of each demographic profile. Health information marketing plans for higher socioeconomic groups 1.1 through 1.3 and lower socioeconomic groups 4.1 through 4.7 will, of necessity, be very different for one reason: Groups 1.1 through 1.3 pay for their own health care, whereas groups 4.1 through 4.7 have no health care or are dependent on social services. In the latter case, marketing strategies would focus more on where social services exist and include promotional information about when and how to use those services. The least a health system can be expected to accomplish is to find fragile populations and create relevant direct marketing of health information to potential consumers.

Another important aid to policy makers is the allocation of health resources. Having discovered fragile populations and created strategic marketing plans, the health system must assure that relevant services exist for health consumers based on their demographic profile. Again, groups 4.1 through 4.7, which exhibit high levels of female-headed households with children younger than eighteen years old, would need different services than groups 2.2 or 3.1, which have an older population. Moreover, allocation of health-system information resources requires centralization and integration, so that the fiscal nature of services, consumers, and providers can be managed to assure conditions of optimal health are achievable.

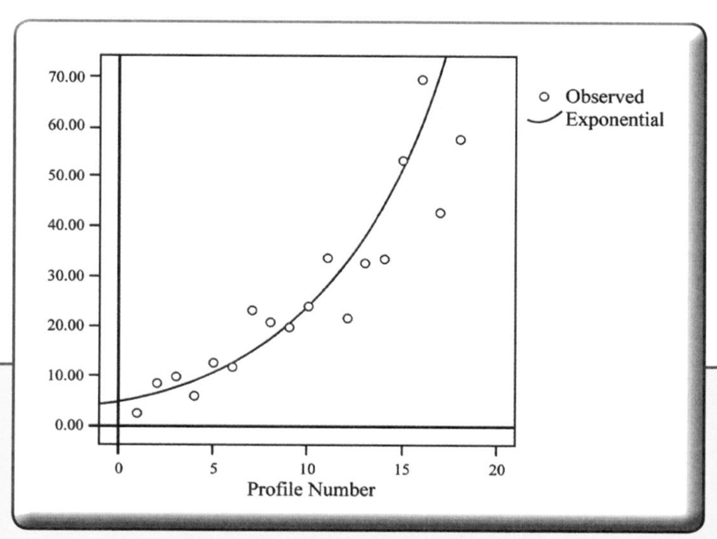

Figure 7.7 The rate of infant abuse has an exponential correlation with demographic profiles.

Discussion

Analysts and policy makers can use the visualization tool, in conjunction with ancillary information, to develop an understanding of the various measures-of-health distributions in relation to the demographic clusters, as well as to target intervention programs in locations where they will be most beneficial to the citizens of Georgia. Programs within DPH that were addressing each measure of health in isolation are now seeing that in many cases those health issues are concentrated in the same demographic clusters. The visualization tool enables a more holistic approach—in other words, seeing "the big picture." A more coordinated effort, prompted by this new tool, will result in more efficient public-health policies and practice.

The ArcReader documents were distributed to the eighteen Georgia health district directors and others involved in the policy-development process. The documents have enhanced the ability of DPH to communicate spatial information regarding the leading causes of YPLL and other measures of health in a timely and efficient manner. Users of the ArcReader documents can freely explore the data in their health districts to determine areas of greatest need. In addition to more efficient distribution of important health information, the ArcReader documents should reduce custom mapping and data requests to the DPH GIS team.

DPH's Children 1st program, in collaboration with the Division of Family and Children Services, is also using the GIS tool to allocate resources more efficiently in the Savannah area. Children 1st "promotes the healthy development of young children and assures that they arrive at school healthy and ready for success." Children 1st is taking into account the relationships between the distribution of infant maltreatment and the demographic profiles to coordinate prevention of child maltreatment. OHIP anticipates that other programs will use the ArcReader visualization tool to further the goal of a safe and healthy Georgia.

Acknowledgments

Millard, F.H. 2005. "Fragile Population Segments in Georgia, USA." Paper presented at "Emerging Issues along Urban-Rural Interfaces: Linking Science and Society," a conference sponsored by the USDA Forest Service, National Science Foundation, and Auburn University's Center for Forest Sustainability, Atlanta, Georgia, March 13-16, 2005.

Enterprise GIS streamlines water-quality assessments

CHAPTER: Supporting policies with GIS
ORGANIZATION: Idaho Department of Environmental Quality
LOCATION: Boise, Idaho
CONTACT: Jim Szpara, senior GIS analyst jim.szpara@deq.idaho.gov
PROJECT: Water-quality assessment
SOFTWARE: ArcGIS, ArcIMS, ArcSDE, ArcView
ROI: Cost and time savings; increased efficiency, accuracy, and productivity; enhanced communication and collaboration; automated workflows; more efficient allocation of resources; improved access to information

By Jim Szpara

With more than 92,000 miles of rivers and streams and more than 1,000 lakes and reservoirs, water is one of Idaho's most important resources. Idaho's rivers, lakes, streams, and wetlands not only provide great natural beauty, but also supply the water necessary for drinking, recreation, industry, agriculture, and aquatic life.

The U.S. Clean Water Act requires states to provide an assessment every two years of the quality of all bodies of water and a list of those that are impaired or threatened. To efficiently meet this charge, the U.S. Environmental Protection Agency (EPA) recommends that states, tribes, and other water-monitoring collaborators use a combination of monitoring and assessment techniques to reliably estimate the overall condition of all waters within a state, assess changes over time, and measure progress toward the "fishable-swimmable" goal of the law.

The Idaho Department of Environmental Quality (IDEQ) is responsible for assuring that Idaho's streams, rivers, lakes, reservoirs, and wetlands meet their designated beneficial uses as well

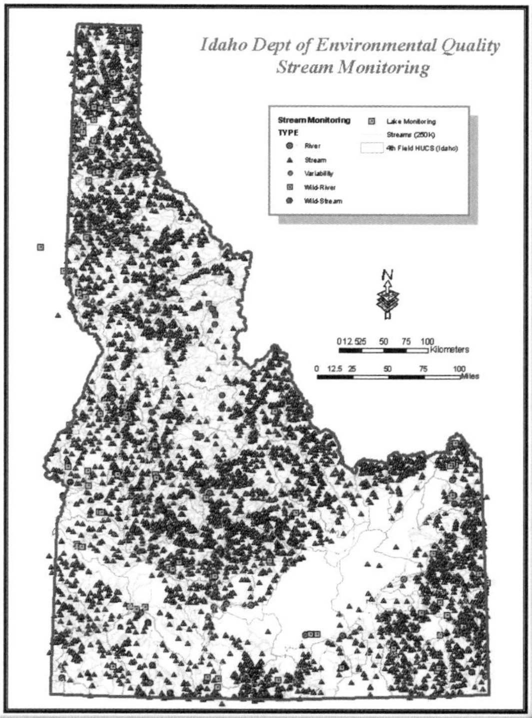

Figure 7.8 IDEQ map shows stream/lake sampling and monitoring locations for 1993 to 2006.

Figure 7.9 IDEQ interactive mapping (ArcIMS) and public-comment Web application displays the results of water-quality assessments and locations of monitoring sites.

as prioritizing site-specific assessments needed to confirm the location of both high-quality and impaired waters. IDEQ must also support control, restoration, and prevention actions under the Clean Water Act.

Since 1993, IDEQ has monitored and assessed nearly 63,000 miles of Idaho's waters, collecting biological, habitat, and chemical samples from more than 7,100 river, stream, and lake locations. To manage and efficiently use the massive amount of water-quality data being collected, IDEQ implemented a statewide enterprise GIS in 2001, which integrates its water-quality monitoring databases with ArcSDE SQL Server and ArcIMS software. Idaho's water-quality standards, or beneficial-use identifiers, were linear referenced to the 100K National Hydrography Dataset (NHD) and used as a primary base layer for analyzing, displaying, and reporting the results of stream water-quality assessments. ArcSDE technology provides indispensable functionality enabling IDEQ to spatially join and relate monitoring and assessment data from multiple SQL server databases to the NHD within ArcSDE. The integrated results are displayed through the Idaho Statewide Area Network with customized ArcView, ArcGIS, and ArcIMS applications to IDEQ's various field offices. They are also available to the public via the Internet with ArcIMS software. This deployment of a centralized enterprise GIS and relational-database solution has increased both the accuracy and efficiency of providing water-quality assessment status reports to the EPA in support of the Clean Water Act.

The Assessment Database (ADB) developed by the Research Triangle Institute (RTI) for the EPA is a relational database application for tracking water-quality assessment data, including use attainment, and causes and sources of impairment. IDEQ's customized deployment of ADB provides water-quality analysts with a user-friendly, intuitive menu-driven process for analyzing monitoring data and reporting water-quality status to the EPA.

Since 1993, IDEQ has monitored and assessed nearly 63,000 miles of Idaho's waters, collecting biological, habitat, and chemical samples from more than 7,100 river, stream, and lake locations.

In-house customizations by IDEQ have seamlessly integrated ADB with ArcGIS and ArcIMS functionality, giving analysts the capability to additionally analyze stream-drainage areas with current land-cover, land-use, and aerial-photography geospatial data from IDEQ's ArcSDE enterprise GIS.

Custom ArcGIS toolbars developed by IDEQ's GIS staff provide "point and click" data automation, enabling analysts and scientists to easily navigate through Idaho's hydrography layers and watersheds while hyperlinked features provide real-time access to water-quality reports and monitoring data.

More information about Idaho's water-quality assessment program is available online at www.deq.idaho.gov/water/prog_issues/surface_water/overview.cfm.

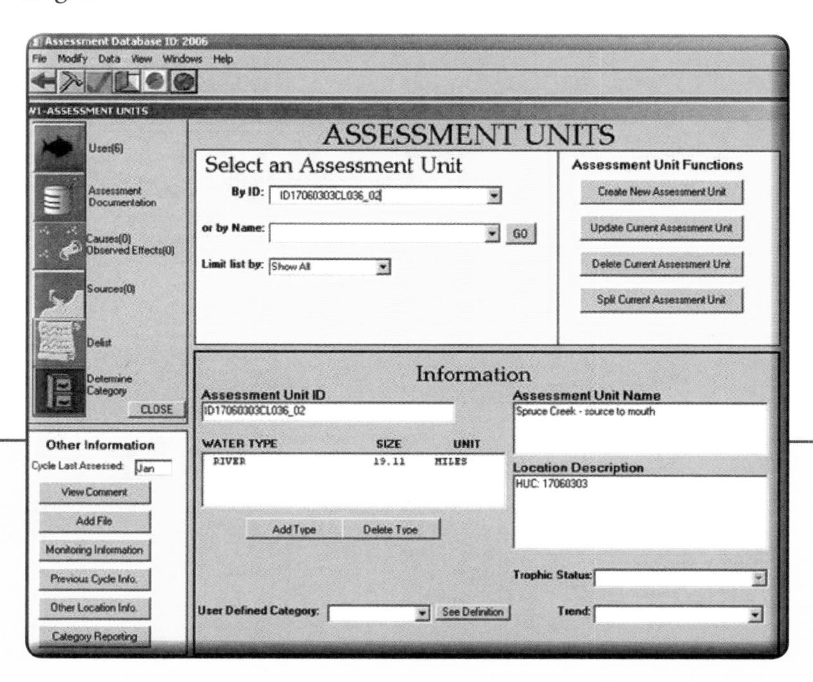

Figure 7.10 The Idaho Department of Environmental Quality's menu-driven assessment database application stores and tracks water-quality assessments.

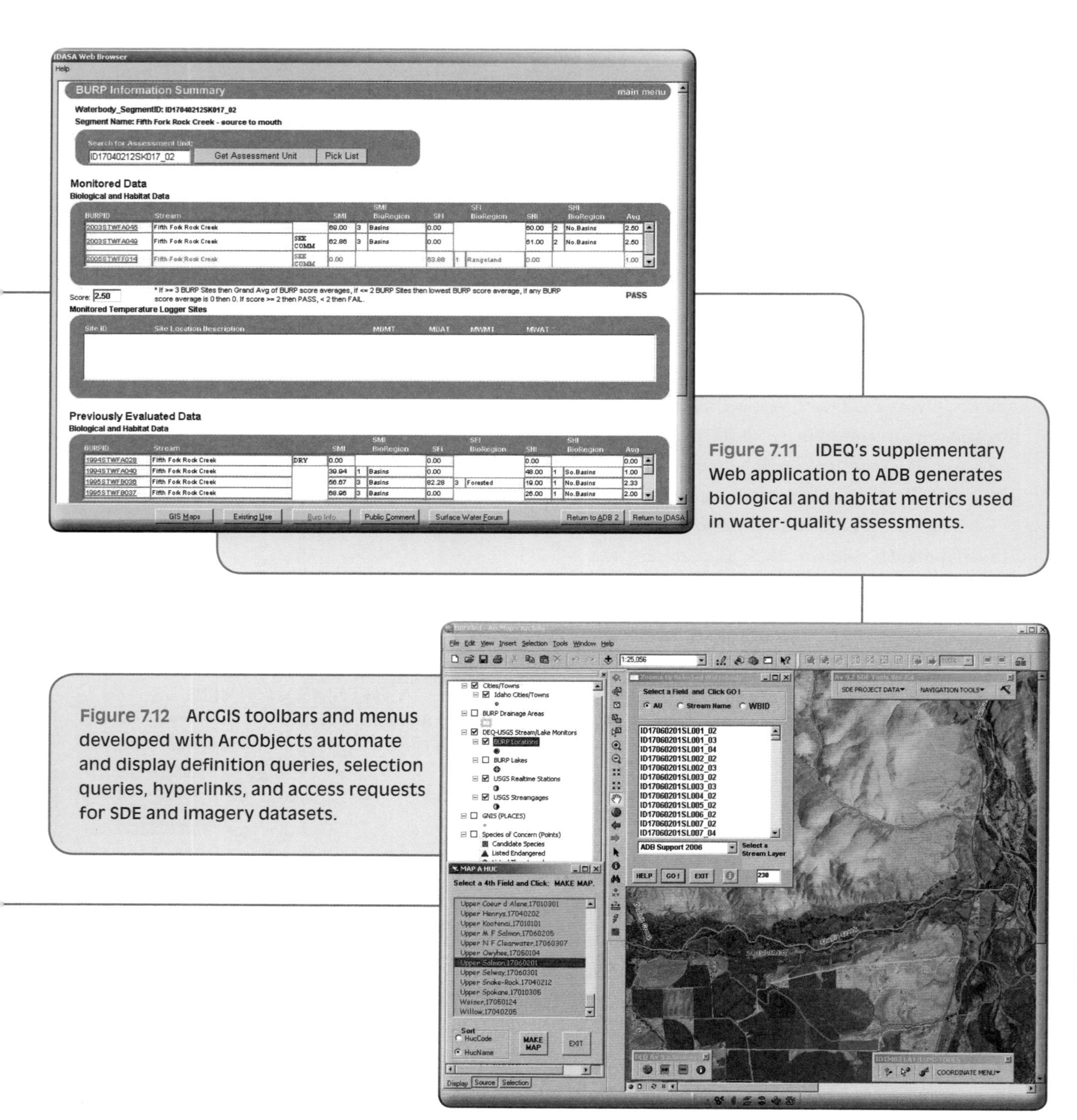

Figure 7.11 IDEQ's supplementary Web application to ADB generates biological and habitat metrics used in water-quality assessments.

Figure 7.12 ArcGIS toolbars and menus developed with ArcObjects automate and display definition queries, selection queries, hyperlinks, and access requests for SDE and imagery datasets.

Preserving open space for the Twin Cities

CHAPTER: Supporting policies with GIS
ORGANIZATION: Metropolitan Council
LOCATION: St. Paul, Minnesota
CONTACT: Rick Gelbmann, GIS manager
rick.gelbmann@met.state.mn.us
PROJECT: Planning for parks and open space
SOFTWARE: ArcView, ArcGIS
ROI: Cost and time savings; increased efficiency, accuracy, and productivity; more efficient allocation of resources; improved access to information

By Rick Gelbmann

The Metropolitan Council is the regional planning agency for the seven-county Minneapolis-St. Paul metropolitan area in Minnesota. In 1967, the Minnesota Legislature established the council to coordinate planning and development within the Twin Cities metropolitan area and to address issues that could not be adequately addressed with existing governmental arrangements. The council oversees the orderly growth of the region, and collects and treats wastewater, operates bus and rail transit, administers the federal Section 8 housing program, and coordinates planning of regional transportation, aviation, water resources, and parks.

The Twin Cities metropolitan area is projected to grow by more than one million people between 2000 and 2030. The Metropolitan Council needed to develop a plan to identify and preserve the best remaining natural resources and outdoor recreational opportunities to meet the needs of the rapidly growing population to 2030 and beyond. The council analyzed a variety of GIS information to determine the following:

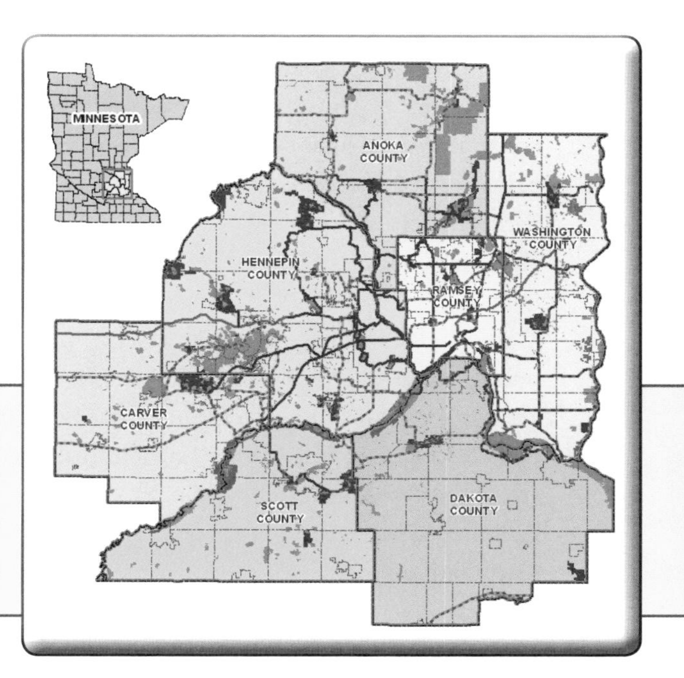

Figure 7.13 The Metropolitan Council for the seven-county Minneapolis-St. Paul metropolitan area plans to add 15,000 acres of regional parks and 196 miles of trails to its current regional system of parks and open space.

- Areas of the region underserved by regional parks in the present and future, based on projected population growth
- Location of the best remaining natural-resources lands
- Costs and potential feasibility of acquiring identified lands

With the detailed analysis in hand and in consultation with local governments in the region, the council developed a visionary plan to add more than 15,000 acres of regional parks and 196 miles of new regional trails to a system that at present encompasses 52,000 acres with forty-eight regional parks, six special recreation features, and 170 miles of trails. GIS mapping made possible both development of the plan and displaying the plan to the public. Implementation of the plan is under way.

Updating a regional parks policy

Minnesota law charges the Metropolitan Council with planning for what are called regional systems, one of which is parks and open space. In 1974, the council designated about 31,000 acres of existing parks owned by the counties, cities, and special park districts as "regional recreation open space." These regional parks had about five million visits in 1975.

By 2004, the regional-parks system open for public use had grown to approximately 50,000 acres and included thirty-five regional parks, eleven park reserves, six special recreation features, and twenty-two regional trails. Annual visits to the regional-parks system had grown to more than thirty million.

Along with adopting a new regional-development framework and growth strategy, the council updated its regional-parks policy plan. The plan contains a variety of policies related to siting, land acquisition, financing, and management of the parks. The 2005 plan also identifies boundary adjustments to existing parks and search areas for new parks and trails to meet the needs of the region's growing population through 2030 and beyond.

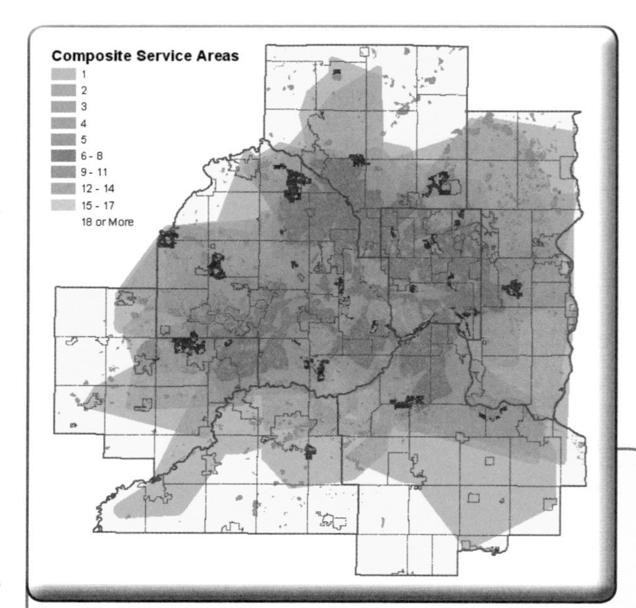

Figure 7.14 Metropolitan Council staff used GIS technology to map out service areas for the region's existing parks.

The availability of robust, accurate GIS data and the capability to analyze the data made identifying and refining the search areas possible. The process that resulted in the new regional-parks system plan includes the following steps:

1. Analyze the service areas of the existing regional parks and identify where gaps exist in the region. The service areas were determined using data collected through in-depth visitor surveys conducted at the regional parks. By combining visitor origin and other data such as transportation analysis zones, staff determined the service area for each park. The data was mapped using GIS technology. When the service areas of all the parks were layered, areas underserved by regional parks emerged. Then the service-area data was merged with data on forecast population growth to determine the areas in greatest need of park facilities.
2. Having determined the broader areas in greatest need of park facilities, the council looked for land within those areas that contained natural-resource features appropriate for preservation and recreational use. The GIS data for this step was provided by the Natural Resources Inventory and Assessment (NRI/A). The NRI/A is a regionwide geographic database that records valuable information about land and water resources that

    - Perform significant ecological functions
    - Contain important habitat for animals that are sensitive to habitat fragmentation and destruction
    - Provide corridors between natural areas to protect the health of plant and animal communities
    - Provide opportunities for people to experience nature and the region's historical landscapes

    In conjunction with the NRI/A data, the council examined parcel data to determine the size of parcels under consideration. A minimum parcel size was established, so that acquisition would be economically feasible. For example, several large parcels next to each other were preferred over two large parcels separated by multiple small ones.

    Park planners made more than sixty individual GIS maps showing potential expansion of existing parks or large natural areas that had no current public-ownership protections. Regional trails were also evaluated or proposed to connect regional parks and preserve high-quality natural corridors. Planners then held meetings with the directors and staff of each of the ten regional-parks-implementing agencies (counties and large cities) to review the maps and get comments. The maps were modified based on local input.
3. Having identified potential parcels, the staff began to make cost estimates of acquisition based on estimated market value contained in a GIS property-tax database. Estimates were made for both potential new parks and for boundary adjustments to existing regional parks. "Without this information, we would have been shooting in the dark as far as estimating what it would cost to pay for the proposed system expansion," said council parks planning analyst Arne Stefferud.

Finally, with well-defined potential search areas in hand, council staff again talked with regional-parks agency staff and elected officials to determine the political feasibility of boundary adjustments to both existing parks and new park sites. "We made adjustments to the plan based on what was happening in the real world," Stefferud said.

Using GIS data and analysis helped the council reduce the cost of parkland acquisition by identifying large land parcels having important natural resources. The use of parcel data and GIS also made it possible to quickly and accurately conduct a comprehensive search for potential parkland and estimate its cost using assessed-value attributes.

The resulting plan identifies the preferred locations for potential parks and trail routes, acknowledging that flexibility will be required over time. A year after the plan was adopted, GIS technician Craig Skone said, "Without GIS, we'd still be working on identifying general search areas." Stefferud said that the parcel and property-tax data made the financial side of the plan "much more credible."

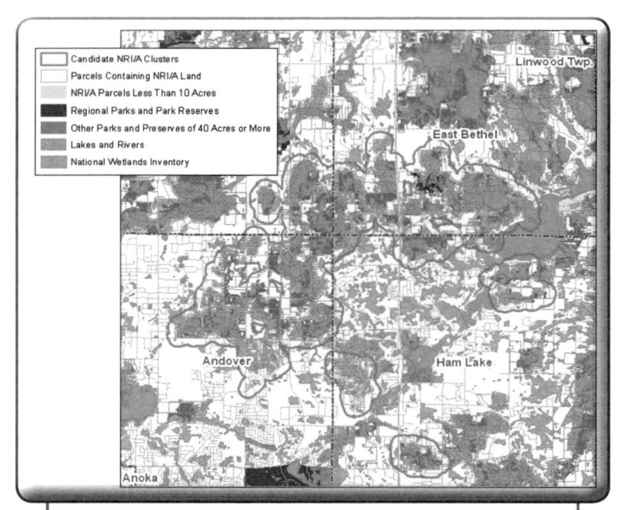

Figure 7.15 Parcels near candidate natural-resource features were mapped for possible acquisition.

> **Without this information, we would have been shooting in the dark as far as estimating what it would cost to pay for the proposed system expansion.**—Arne Stefferud, council parks planning analyst

Citizen response to the council's visionary parks plan in a series of public hearings was extremely positive. A map of the plan gives people a clear visual representation of the council's future vision for parks. "This is the best parks plan the council has created since the 1970s when the system was first established," said John Herman, a Minneapolis attorney and conservationist who lobbied for passage of the legislation that created the regional-parks system in 1974. "The plan lays out a vision for protecting some of the few natural areas remaining in the region. I think GIS mapping is a fantastic tool to help policy makers envision what is possible."

The council uses a PC-based network of computers that run on Microsoft operating systems. The project was started using ESRI's ArcView and was completed with ArcGIS. Two parks planners and a GIS technician completed the in-depth parks system analysis that included development of the parks service areas. The development of the parks plan itself involved a parks planner, a planning analyst, and a GIS technician.

Using GIS to assess preservation legislation in New York

CHAPTER: Supporting policies with GIS
ORGANIZATION: Preservation League of New York State
LOCATION: Albany, New York
CONTACT: John Knoerl, PhD, president and chief cartographer, KEI Maps Inc. keimaps@comcast.net
Daniel Mackay, director of public policy, Preservation League of New York State
PROJECT: Support preservation legislation
SOFTWARE: ArcExplorer, ArcGIS
ROI: Increased efficiency, accuracy, and productivity; improved access to information

By Daniel Mackay and John J. Knoerl

GIS technology is most often thought of as a tool for managing geographical resources at the state and local levels. Business, real estate, health-services, education, and natural- and cultural-resource management applications are commonplace and are being used to good effect. The use of GIS as a tool in the arena of public policy and legislation, however, is not as widespread as these more traditional applications. Legislators, their staff, advocates, and opponents rarely use GIS to analyze the impact of preservation legislation. Yet many parts of preservation laws are replete with spatial provisions. If GIS were used to evaluate or predict the intended—as well as the unintended—effects of such proposed laws, then the legislative process would be well served, and possibly better legislation and preservation would result.

In support of a continuing campaign to secure a New York state income tax credit for rehabilitation of historic residential structures, the Preservation League of New York State (PLNYS) launched a GIS-based initiative to quantitatively assess the impacts of two tax credit proposals before the New York State Legislature. A systematic comparison of the impacts of these legislative proposals was required to focus attention and generate additional legislative support for the program. With partial funding secured from the National Trust for Historic Preservation, the league employed the services of KEI Maps to create and manage the GIS database at the core of this work.

Such a comparison would have been impossible using anecdotal data. However, by employing GIS as a tool, quantitative data about who would benefit from each proposal was clearly determined and brought into the debate. GIS analyses elevated the discussion from the realm of anecdotes and speculation to the quantitative. In turn, this quantitative analysis provided critical information daily that informed the legislative outreach conducted by the league.

PLNYS has long been involved with advocating governmental policy that is sensitive to and supportive of historic preservation. Through advocacy, technical assistance, grants, and support, the PLNYS has successfully changed governmental behavior at both the statewide and local levels.

GIS analyses elevated the discussion from the realm of anecdotes and speculation to the quantitative.

Increasingly, the league has come to realize that to be successful in its advocacy role, it must develop a strategy that relies on the use of comprehensive data as opposed to anecdotal data to support its policy agenda. The critical importance of the proposed New York State Neighborhood Reinvestment Act provided an excellent opportunity to begin to develop a comprehensive database of information about historic properties in New York State that could be used to argue for passage of the legislation.

As it was, the governor had an alternative version of the bill he had introduced to the New York State Assembly. The comprehensive database was immediately put to use to draw comparisons between the two versions of the bill. This facilitated discussions that were more detailed and meaningful. Both advocates, and legislators and the governor's staff were able to refer to the same set of information. Tables and maps were generated in what may be best described as real time given the fast-paced setting of the legislative process.

Building the database

The project was organized into four phases: data acquisition, data preparation, data analysis, and reportage.

1. Data acquisition—Data acquisition involved establishing the objectives of the analyses, identifying the data needed to conduct the analyses, identifying the sources for acquiring the data, and acquiring the data itself.

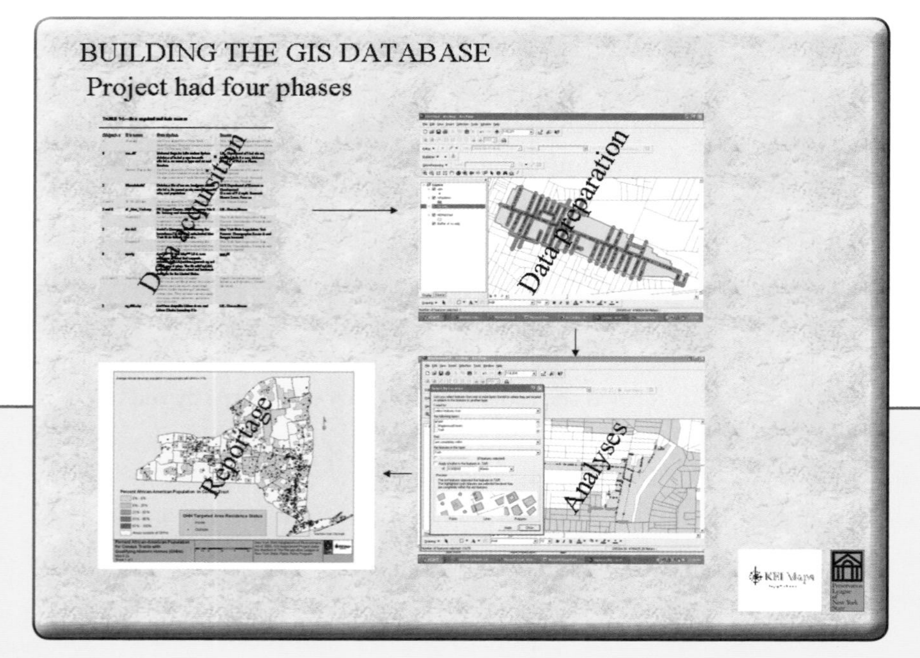

Figure 7.16 The Preservation League of New York State used four phases to create a GIS database of homes in the state's historic districts.

The three objectives of the analyses were

- To determine the number of potentially qualifying historic homes (QHH) for each version of the New York State Neighborhood Reinvestment Act of 2003
- To determine where these potential QHH were located with respect to targeted area residences, state empire zones, zones of equivalent areas, qualified census tracts, urban areas, New York State Assembly districts, New York State Senate districts, New York State U.S. congressional districts, New York State Democratic Assembly districts, and New York State Republican Assembly districts
- To develop a profile of the neighborhoods within which these potential QHH are located by using Census 2000 housing variables of percentage of home ownership, racial and ethnic diversity, average median value of housing units, average median monthly owner costs of housing units, and median family income

2. Data preparation—The raw data received from the various sources was examined, projected into Universal Transverse Mercator (UTM) Zone 18 using North American Datum 1983, merged with stand-alone database tables, and in certain instances, manipulated to achieve the appropriate data categories for each analysis.

There were two parameters used in creating the QHH.shp layer:

- Estimating the number of contributing residential properties in historic districts
- Locating the contributing residential properties within their historic districts

The map layer of historic properties provided by the New York State Historic Preservation Office did not include the number of contributing

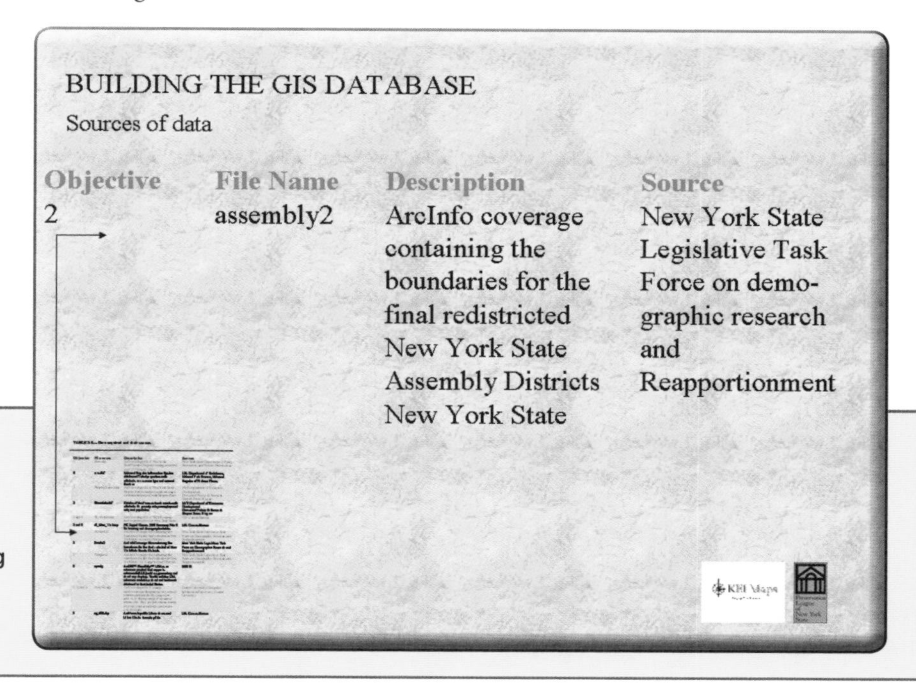

Figure 7.17 The league had to first set objectives for building its GIS database.

properties within each historic district. By joining the layer to an Access database of properties listed on the National Register of Historic Places, the number was obtained, but it was not broken down by building type—residential, commercial, or industrial. To estimate the number of residential buildings in historic districts, a statistically valid random sample of eighteen historic districts in New York State was drawn from a population of 488 historic districts coded as residential. An average of 76 percent of contributing properties in the sample districts was residential. The remaining properties were churches, schools, and some commercial structures. The total number of contributing properties in a district was multiplied by 0.76 to get the number of residential properties in each historic district.

The locations of contributing properties within a historic district were not always mapped. However, it was important to make an estimate of where the contributing properties were located to determine if they were in a targeted district residence. Two procedures were used for this estimate:

- Randomly place each contributing property within the district but constrained to a buffer within 10 feet of a street in the district
- Use address matching or geocoding if the addresses were listed in the record of the historic district

3. Data analysis—Using the GIS database, which included census-tract data, many important questions were answered. For example, how many historic homes would qualify under the proposed legislation? How many of these homes are located in each assembly, senate, and congressional district? How many structures are located in urban and rural areas? How many qualified homes are owner-occupied? What is the racial and ethnic diversity of neighborhoods with

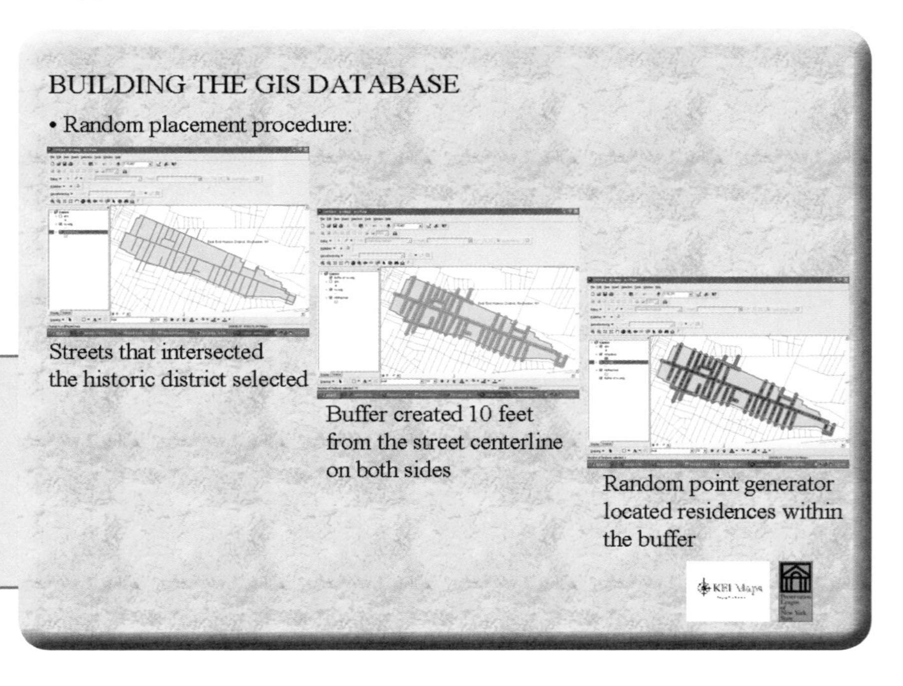

Figure 7.18 The league used a 10-foot buffer to determine potentially qualifying historic homes.

QHH? What is the average median value of homes in neighborhoods with QHH? What are the median monthly mortgage costs, and what is the median family income for neighborhoods with QHH? Creating this database enabled the league to peer into the demographic heart and soul of New York State's more than 700 National Register Districts for the first time and provided the league with a unique tool to inform and advance preservation policy.

4. Reportage—Twenty-nine tables were generated for the report. These tables were started in ArcMap using the report tool to create the initial format. Ultimately, they were brought into the Word format and then converted into a table for final editing and formatting.

Twenty-four maps were generated for the report. A template was developed and used consistently in the generation of the maps. In numerous instances, an inset map of the New York City area was required in order to display the data in that area. The maps were formatted in ArcMap and exported as a PDF image.

A CD was created containing the following items:

- The New York State Neighborhood Reinvestment Act of 2003: A Geographic Information System Assessment as a Word document
- Map images in PDF
- Tables as rich text files
- Shapefiles: QHH, NY Historic Properties, TAR, SEZ, ZEA, QCT, Assembly, Senate, Congress, urban, tr36_utm00, ALLNR, counties
- DBF files: DIVERSIT, HOUSEINC, HOUSEVAL, MEDFAMIN, OWNERSHI, OWNERCOS
- ArcExplorer GIS viewer software to view the shapefiles and DBF files

Creating this database enabled the league to peer into the demographic heart and soul of New York State's more than 700 National Register Districts for the first time and provided the league with a unique tool to inform and advance preservation policy.

Improvements and future uses

Given the initial success of the GIS database in the context of the New York State Neighborhood Reinvestment Act what is PLNYS' vision for improving and using it in the future? The immediate improvement is to accurately locate the contributing historic properties, both residential and commercial, within the state's historic districts. One potentially promising approach is to link the locations of contributing historic buildings to the New York State

Office of Real Property's GIS database. The database has a point for every tax parcel for most of the counties in New York State and also contains the street number and name for each tax-parcel point. The latter data can be used to match up the street addresses of the contributing properties. This data is likely to be accurate to the extent that it is based on tax-parcel maps, which are usually drawn to a 1:2400 scale or better.

Not only will the locations of contributing properties be more accurate, but by linking them to the real-property database, all of the information contained in that database will now also become part of the GIS historic-properties database. Information such as the assessed value of the property can be used by PLNYS to assess the dynamics of change in historic neighborhoods, for example. Other types of information that can be incorporated in the GIS database include the date the house was built, ownership of the property, property code (e.g., single-family, two-family dwelling), and tax parcel ID.

The second priority for improving the database is to add data on locally designated properties and historic districts. This data is key to developing and advocating preservation policy that is relevant to the local level of historic preservation.

Third, the database can also be used to model the impact of the national Historic Homeowners Assistance Act on historic properties in New York State. The bill was reintroduced in the U.S. Congress. This assessment could prove vital to convincing the New York State congressional delegation to support such legislation. Currently, support is not strong among the New York delegation.

Fourth, each year the PLNYS selects seven historic places to save. The "Seven to

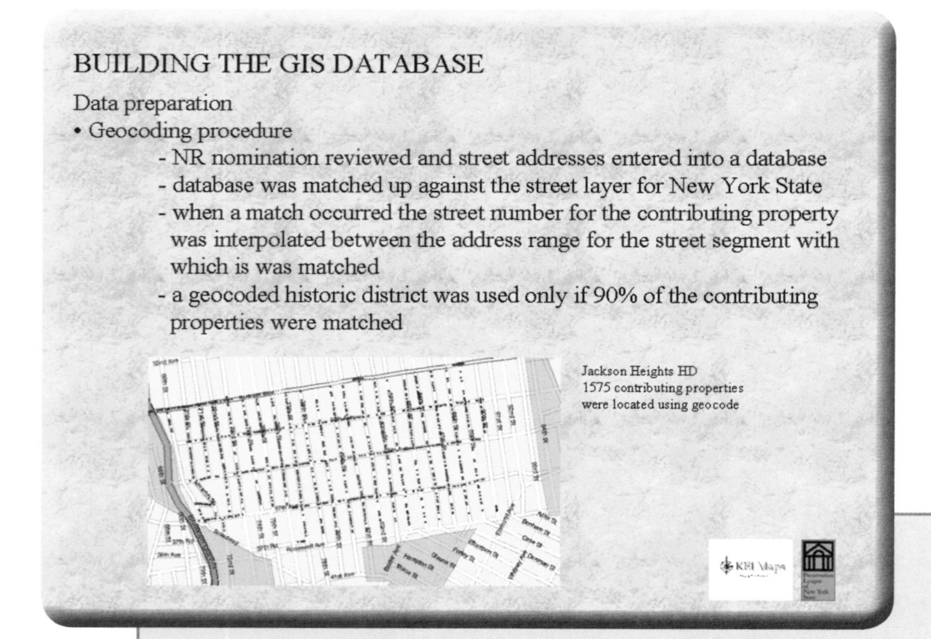

Figure 7.19 Geocoding was one of the methods used to determine where qualifying historic homes were located.

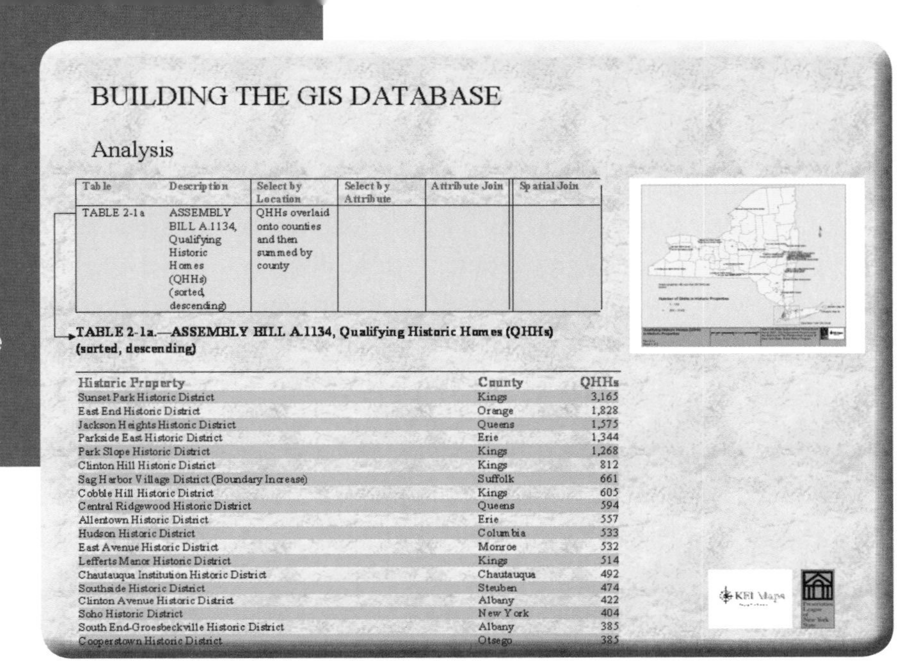

Table	Description	Select by Location	Select by Attribute	Attribute Join	Spatial Join
TABLE 2-1a	ASSEMBLY BILL A.1134, Qualifying Historic Homes (QHHs) (sorted, descending)	QHHs overlaid onto counties and then summed by county			

TABLE 2-1a.—ASSEMBLY BILL A.1134, Qualifying Historic Homes (QHHs) (sorted, descending)

Historic Property	County	QHHs
Sunset Park Historic District	Kings	3,165
East End Historic District	Orange	1,828
Jackson Heights Historic District	Queens	1,575
Parkside East Historic District	Erie	1,344
Park Slope Historic District	Kings	1,268
Clinton Hill Historic District	Kings	812
Sag Harbor Village District (Boundary Increase)	Suffolk	661
Cobble Hill Historic District	Kings	605
Central Ridgewood Historic District	Queens	594
Allentown Historic District	Erie	557
Hudson Historic District	Columbia	533
East Avenue Historic District	Monroe	532
Lefferts Manor Historic District	Kings	514
Chautauqua Institution Historic District	Chautauqua	492
Southside Historic District	Steuben	474
Clinton Avenue Historic District	Albany	422
Soho Historic District	New York	404
South End-Groesbeckville Historic District	Albany	385
Cooperstown Historic District	Otsego	385

Figure 7.20 The GIS database allowed the league to see how many residential homes were in each historic district.

Save” program draws attention to important preservation issues such as abandonment, disinvestment, demolition, and development through the selection of seven properties that reflect these problems. The GIS database could play an important role in making the selection with an eye toward the geographic distribution of these selected properties. It would also be possible to use the GIS database to extend the concept to saving entire historic neighborhoods. The database could be used to identify

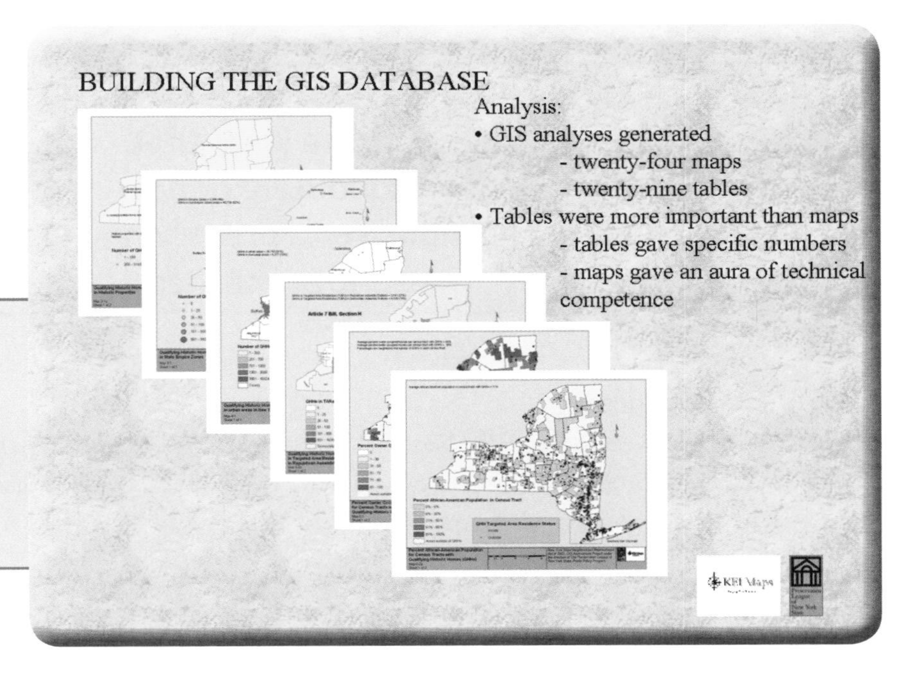

Figure 7.21 The GIS analysis generated an assortment of relevant maps and tables.

potential candidate neighborhoods using census data that reflects the demographics and housing variables. Census data can be tied to preservation issues such as gentrification, ethnic diversity, accelerated ownership turnover, housing values, and ownership costs. The historic neighborhoods selected to save could be a rallying point for local leaders and residents alike.

Fifth, for the long term, the PLNYS could use this database to set up a monitoring program across the state. Such a program would continually monitor the vital signs or health of historic properties and provide early warning on developments that could be potentially detrimental to the preservation of these properties. Additionally, summary statistics could be generated quickly when needed as well as annually, such as an end-of-year report on the state of historic properties. The database could be used to establish criteria that would define areas across the state that are conducive or not conducive to preserving historic properties. Such targeted areas could be prioritized for PLNYS intervention, thus making the most of available preservation funds.

In advocating the passage of the New York State Neighborhood Reinvestment Act the PLNYS believed that GIS could be a useful tool in arguing the merits of this legislation. With spatial overlays, quantitative data on who would benefit from the bill's passage was clearly determined and brought into the debate. PLNYS is confident its situation is not unique and that other preservation advocacy groups could benefit from using GIS in similar ways.

The database could be used to establish criteria that would define areas across the state that are conducive or not conducive to preserving historic properties.

Acknowledgments

The authors wish to thank the National Trust for Historic Preservation for providing partial financial support for the project. The genesis of using GIS to model legislation began with its inquiry on the number of historic homes that might qualify under the Historic Homeowners Assistance Act of 2002. Its support is gratefully acknowledged.

Maintaining downtown curb appeal in Sheridan, Wyoming

CHAPTER: Supporting policies with GIS
ORGANIZATION: City of Sheridan Planning Division
LOCATION: Sheridan, Wyoming
CONTACT: Scott Lieske, research scientist, University of Wyoming lieske@uwyo.edu
PROJECT: Assessing parking requirements
SOFTWARE: ArcView
ROI: Increased efficiency, accuracy, and productivity

By Scott Lieske

Compact walkable blocks with active storefronts often define the sense of place along Main Street in small Western towns. Sheridan, Wyoming, is a classic example with its Main Street Historic District serving as the center of community and business activity against a backdrop of late nineteenth- and early twentieth-century architecture.

Parking can have a tremendous impact on the look and feel of a community, and often, the traditional requirements for minimum-parking provisions in zoning codes result in large surface parking lots that tend to separate uses and deaden downtowns. In Sheridan and similar Main Street-focused communities, it is important to evaluate whether these parking requirements will enhance or detract from the qualities that make these areas so special.

The City of Sheridan's goals for its historic district, as stated in the Vision 2020: Sheridan County Growth Management Plan, include maintaining "a community character that preserves the quality of life, values, and traditions of the area" as well as maintaining "the downtown as a primary hub of the community, to minimize through traffic, and to support the downtown as both a commercial center and gathering area." Implementation strategies in support of these goals include encouraging efforts to maintain and improve the historic character of the downtown and using capital improvements to promote a pedestrian-friendly downtown. Given downtown Sheridan's existing vibrant character,

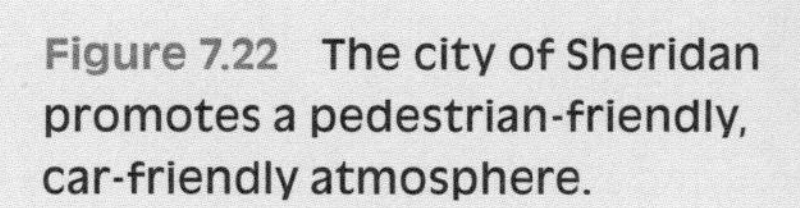

Figure 7.22 The city of Sheridan promotes a pedestrian-friendly, car-friendly atmosphere.

the desire to maintain it, and the potential impact of parking on the area, citizens and business groups chose to review existing parking requirements to determine whether they supported the desires of the community in preserving the flavor of the Main Street Historic District.

Assessing parking policies

The overall goal of the project was to quantitatively assess the city's parking requirements. Parking regulations were evaluated by comparing them with different sets of minimum-parking standards, comparing the standards with existing parking supply, and relating parking requirements and supply to existing land use. The project also demonstrated the use of GIS-based planning decision support system tools in providing a cost-effective assessment of parking issues in a municipality and improved access to information about downtown parking to citizens and decision makers. The analysis allowed a detailed overview of parking requirements and the current parking situation without the expense of a detailed parking survey. Broader impacts of the project suggested a reversal of the conventional wisdom that national standards for minimum-parking requirements will maintain the vitality and character of downtown Sheridan and, indirectly, assisted city staff with implementation of the Sheridan County Growth Management Plan.

Project partners included the City of Sheridan's Planning and GIS divisions and the University of Wyoming. Resources were brought to bear with the assistance of the Plan-IT Wyoming Partnership, which coordinates efforts between the University of Wyoming and Wyoming communities to build the capacity to benefit from GIS and planning support tools. In Sheridan, the project drew upon the expertise of Planner Robert Briggs, planning technician Meg Poulson, and GIS Coordinator Steve Lowman. At the university, Phil Polzer, research scientist with the Wyoming Geographic Information Science Center, contributed to the project.

The analysis allowed a detailed overview of parking requirements and the current parking situation without the expense of a detailed parking survey.

The initial objective of the modeling effort was to produce a quantitative comparison of minimum-parking standards. The university compiled three sets of parking standards in tabular format: the City of Sheridan's parking requirements, the standards published by the Institute of Transportation Engineers (ITE), and the standards of the American Planning Association (APA). Data was recorded as the number of spaces required per 1,000 square feet of building floor area. Because the standards are all based on land use, requirements were linked in the table to land-use designations from the American Planning Association's Land Based Classification Standards (LBCS) activity codes.

The second modeling objective was to compare required-parking specifications with available parking (defined as parking within a 300-foot distance of each structure). This entailed developing detailed data layers for both parking supply and land use. The parking-supply layer was developed by Poulson and completed in late 2005. The layer shows precise locations for parking spots and includes attributes for (a) type of parking, including fifteen-minute, twenty-minute, two-hour, handicapped, private, and unrestricted; (b) type of surface; (c) markings; and (d) space size. The land-use layer was derived from a layer of building footprints suitable for large-scale analysis. To obtain a coarse assessment of land use, building footprints were first attributed with parcel-level data from the Sheridan County assessor's database. This data was expanded and augmented through a windshield-level survey that added land-use data for each floor of each building using LBCS activity codes. Also added were the necessary z values so building footprints could be properly placed and extruded in a 3D environment.

The next step in comparing required parking with available parking was to use planning support tools to relate the data layers to one another and analyze resulting outcomes. CommunityViz software (Placeways LLC; Boulder, Colorado) was implemented to calculate minimum-parking requirements for each land-use record by determining the area for each record and multiplying that area by the corresponding value (linked by the LBCS activity codes) from the minimum-parking requirements table. Parking requirements were then stored as attributes in the land-use layer.

Relating parking requirements to parking supply is an iterative process, and Polzer developed a customized parking-allocation application to perform the analysis. The application relates land-use and parking requirements to available parking distributions (stored in a separate layer), enabling the user to select the land-use layer, an allocation field, the parking-spaces layer, and a minimum-parking-requirements table. After selecting the data inputs, the user specifies values for "maxdistance," allocation process, allocation method, and allocation type.

Maxdistance is the maximum distance a parking space can be from a building and still be considered available to the building. Allocation process is

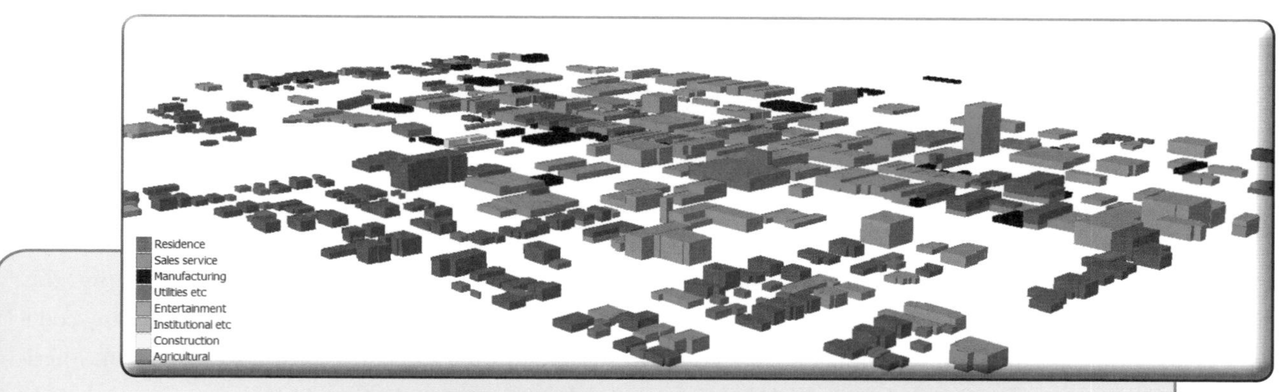

Figure 7.23 Building footprints for downtown Sheridan, Wyoming, are extruded in a 3D environment.

a choice between allocating parking spaces up to the point that buildings meet their parking requirements or until all available parking spaces are used. Allocation method is a choice between using the closest available parking space or the first available space in the attribute table of the parking layer. Choices for allocation type are "one time through" or "iterative." The Parking Allocation Tool stores results in a designated field in the land-use layer.

The impact analysis capabilities of CommunityViz were next used to develop and chart indicators of parking supply and demand. The chart, "Parking Summary," aggregates overall parking supply and demand in the Main Street Historic District. Parking supply is the total number of spaces available regardless of type (4,951). Also in the chart are summations of total calculated demand based on the three sets of parking standards (APA, ITE, and the city's). These indicators simply add the relevant demand field (APA, ITE, or Sheridan) from the appropriate land-use layer while excluding vacant properties. The chart also shows parking demand as calculated by the Parking Allocation Tool. This data is a summation of the number of spots allocated to each record and is presented in the chart in the columns APA_tool, ITE_tool, and Sheridan_tool.

The analysis

Parking demand based on the three sets of minimum-parking standards was in all cases significantly higher than existing supply. The similarity between the overall demand calculated for the three sets of standards and the difference between overall calculated demand and the availability of spaces suggests there should be a severe shortage of parking in downtown Sheridan. Anecdotal evidence, however, suggests otherwise. With the exception of extremely busy times such as during a summer festival or rodeo, it is almost never difficult to find a parking spot within a couple of blocks of your destination. Overall numbers for parking supply and demand suggest parking requirements in the City of Sheridan's zoning code and the alternative national standards are too high.

The Parking Allocation Tool allows for a more precise analysis by moving the scale of the evaluation from the entire Sheridan Main Street Historic District to the level of each land-use record. Results were summarized statistically and viewed spatially. Using the APA parking standards, 69 percent of buildings meet their parking requirements; under ITE parking standards, 76 percent of buildings meet requirements; and under city requirements, 67 percent of buildings meet requirements. Buildings that meet their requirement are almost entirely those with low parking requirements. The majority (199 of 283) of buildings with a demand of greater than six spaces fall short of their required number of spaces. The spatial distribution of buildings that do not meet requirements is also telling. Under all three parking-demand scenarios, the majority of smaller buildings that don't meet requirements are on Main Street. Under the city's parking requirements, 143 of 189 land-use records that don't meet requirements are on Main Street.

Minimum-parking requirements set by the ITE, APA, and the City of Sheridan's zoning code are too high for a mixed-use area such as the Sheridan Main Street Historic District. Coupled with

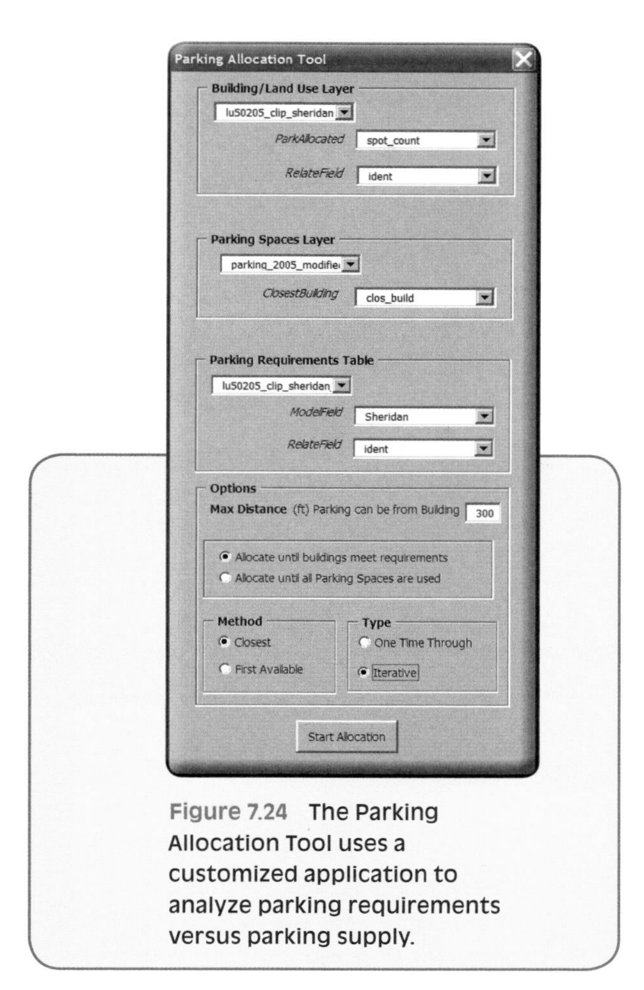

Figure 7.24 The Parking Allocation Tool uses a customized application to analyze parking requirements versus parking supply.

anecdotal evidence that the supply of parking is adequate, general results from parking supply versus calculated demand, and the results of Parking Allocation Tool runs suggest that some percentage of overall parking requirements will be sufficient to meet demand in the historic district. The idea that minimum-parking requirements can be reduced in a mixed-use area is well supported in the literature. In mixed-use areas, demand for parking is reduced, and consequently parking requirements can be reduced, because people are able to park once and visit several establishments. A 2004 Montana study found that central business districts require only 60 percent of mandated parking.

There are a number of policy options that modify or provide alternatives to minimum-parking requirements. A municipality can simply lower requirements in mixed-use districts and areas well served by transit or in any district simply to match real demand. Another option is to charge developers a fee in lieu of providing required parking. The fee is paid per space to a municipality in compensation for not providing required parking. The municipality can then use the funds to provide

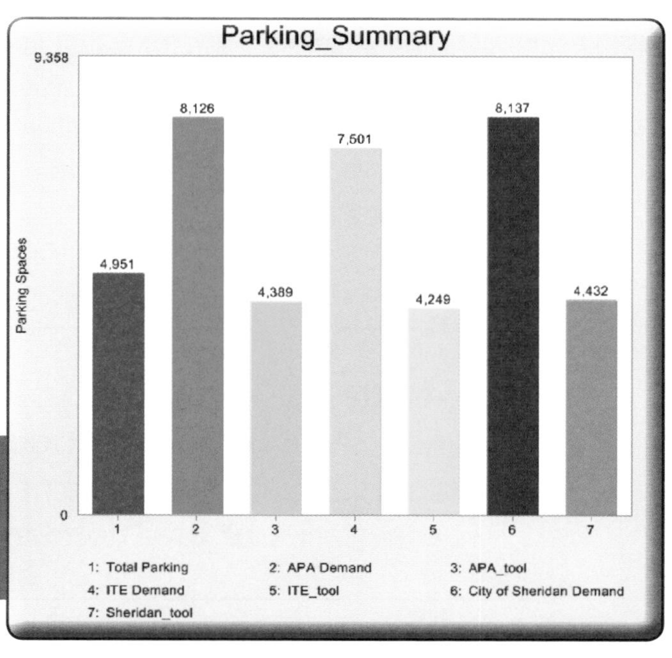

Figure 7.25 The Parking Summary compares parking supply with various parking standards.

parking that serves several businesses. This option provides flexibility to businesses, especially in high-land-value central business district locations. Market-rate parking can be used to efficiently allocate parking and is politically palatable when revenues are spent on local improvements, often through a parking benefit district.

This study suggests the city's parking requirements are not likely to help realize its goals for the historic district. Excessive minimum-parking requirements have negative impacts including separation of uses and deadening of downtowns that are contrary to the goals expressed in the Sheridan County Growth Management Plan. Over time, if these parking requirements are maintained, they will likely degrade the quality of the built environment and decrease overall vitality in downtown Sheridan. On the other hand, a well-developed understanding of the relationship between parking and the built environment will facilitate growth and change, ensure adequate parking is available, and enhance the quality of the Sheridan Main Street Historic District.

Figure 7.26 Downtown Sheridan doesn't let parking get in the way of its historic flair.

Acknowledgments

The author thanks Jeff Hamerlinck, director of the Wyoming Geographic Information Science Center, for comments and suggestions. This project was supported by the Plan-IT Wyoming Initiative, with funds from the Wyoming Business Council, Qwest Communications, the Wyoming Community Foundation, and the University of Wyoming's Ruckelshaus Institute of Environment and Natural Resources, and Wyoming Geographic Information Science Center.

A well-developed understanding of the relationship between parking and the built environment will facilitate growth and change, ensure adequate parking is available, and enhance the quality of the Sheridan Main Street Historic District.

A Web-based GIS tool for railroad hazmat routing

CHAPTER: Supporting policies with GIS
ORGANIZATION: Oak Ridge National Laboratory and University of Tennessee
LOCATION: Knoxville, Tennessee
CONTACT: Shih-Miao Chin, PhD, civil engineer chins@ornl.gov
PROJECT: Tool for hazmat routing
SOFTWARE: ArcIMS
ROI: Cost and time savings; increased efficiency, accuracy, and productivity; enhanced communication and collaboration

By Shih-Miao Chin

Vigilant government jurisdictions understand the risks of having hazardous materials (hazmat) move through densely populated areas and have begun to take preventive actions. A hazmat shipment ban legislated in Washington, D.C., led to debates, legal challenges, and consideration by other major cities of whether to pursue similar actions.

There is no indication from previous railroad incidents that there are frequent hazmat mishaps. Data from 1990 through 2002 show an average of twenty-five hazmat transport incidents (crashes or derailments) per month with the trend slightly increasing over time. On the other hand, the timely and reliable supply of these chemical and petroleum products/materials are of vital importance to the economic health of the nation. While the constitutional legality of such a unilateral shipping ban by a city government has been contested, some in the transportation industry also believe that rerouting a sizable number of shipments, hazmat or not, could potentially incur much higher transportation and inventory costs without necessarily improving the safety and security of the public.

Figure 7.27 CSX Transportation track runs through the heart of Washington, D.C.

Framework for routing

The tool for hazmat routing evaluation and alternative transportation, or THREAT, is a Web-based GIS tool for routing hazmat shipments via railroad networks under scenarios such as shipment bans. THREAT is capable of searching for the best routes that optimize prescribed objective functions, calculating an array of performance and operational measures for each route, comparing different routing alternatives, and generating animated routing maps. This tool helps transportation analysts, policy makers, and officials to evaluate alternatives easily and quickly, then effectively communicate their decisions to the public. The legal merits of the issues raised by the two sides of the ban are not discussed, but rather, an evaluation tool is presented to aid the assessment of routing alternatives based on the risks of exposure and the general costs of operations.

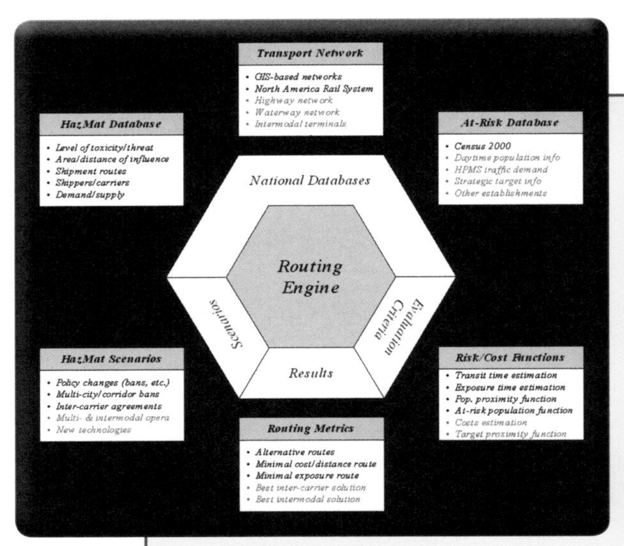

Figure 7.28 A variety of considerations went into the conceptual framework for a hazmat-routing evaluation.

This tool helps transportation analysts, policy makers, and officials to evaluate alternatives easily and quickly, then effectively communicate their decisions to the public.

To assess the risks and costs of various routing scenarios when access to a portion of the overall network is denied, an analysis framework was conceptualized. This framework employs a collection of national databases, considers selected shipment constraints and scenarios, estimates the risks and costs of all routing alternatives, and generates a number of statistical measures that aim to aid policy or operational decision making. This software tool runs on a Windows-based server.

The study

To demonstrate the use of THREAT, a case study based on the hazmat shipment ban in Washington, D.C., was conducted. CSX Transportation, which owns the largest rail network in the eastern United States, covering 23,000 track-miles in twenty-three states and two Canadian provinces, was the hazmat carrier for the study. The goal was to assess and identify suitable alternative routes that would circumvent Washington, D.C. For the estimation of population at risk, a three-mile-wide exposure zone, with a 1.5-mile radius from any point, along CSX Transportation tracks was designated for the measure of risk. Population-density data in the exposure zone is available from Census 2000.

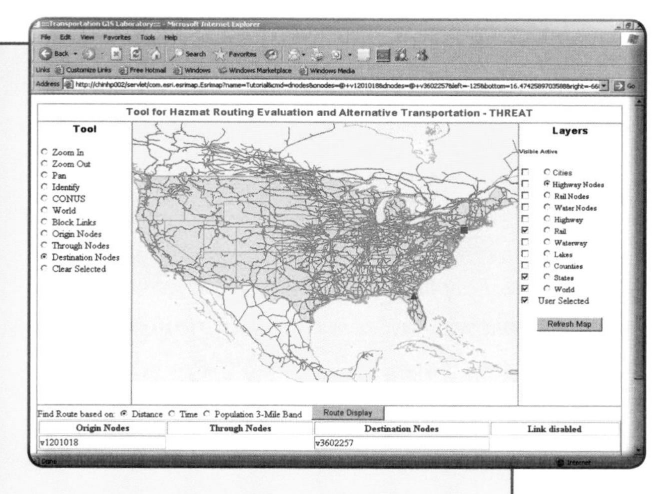

Figure 7.29 THREAT is a Web-based GIS tool for hazmat routing.

Study scenarios

To simplify the magnitude and complexity of the case study while still demonstrating how THREAT works, results for only one selected pair of origin and destination (OD) points, Jacksonville, Florida, and Jersey City, New Jersey, are presented here. With the hazardous-shipment ban in Washington, D.C., the study investigated four routing scenarios:

- *Base case*—The status quo before the shipment ban in Washington, D.C. The base case considers no shipment restriction around Washington, D.C., and the route follows mostly the I-95 corridor. A shipment on this route would go through cities such as Savannah, Georgia; Charleston, South Carolina; Fayetteville, North Carolina; Richmond, Virginia; Washington, D.C.; Baltimore, Maryland; and Philadelphia, Pennsylvania, to reach its destination in Jersey City, New Jersey.
- *Shortest-distance alternative*—The shortest-distance route on CSX Transportation tracks without passing through Washington, D.C. This case considers the shortest route between Jacksonville and Jersey City, while avoiding the Washington, D.C., Capitol exclusion zone (CEZ). It shares part of the same route, between Jacksonville and Savannah, with the base case before diverting west through Augusta, Georgia; Spartanburg, South Carolina; and Kingsport, Tennessee. A large portion of this route travels through the Appalachian Mountains. It rejoins the base-case route near Baltimore and continues through Philadelphia to reach Jersey City. Because of the mountainous terrain and coal-train traffic in the Appalachians, this alternative, although the shortest in distance, is an unlikely candidate for a hazmat shipment.

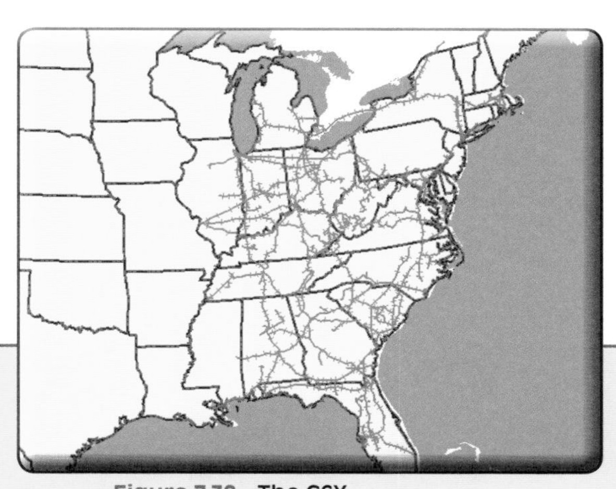

Figure 7.30 The CSX Transportation railroad network is analyzed for routing hazmat.

- *Most-likely alternative*—Railroad companies prefer to ship hazmat on high-quality tracks that are frequently maintained to minimize the likelihood of mishaps resulting from equipment failure. However, a competing objective of railroad companies is to stay profitable and, hence, not to overmaintain their tracks. The most likely route will use these high-quality main-line tracks—a subset of the CSX Transportation network. Using only CSX Transportation's high-quality main-line tracks while avoiding the Washington, D.C., CEZ, the route would be diverted to the west immediately after crossing the Georgia state line, passing through the cities of Atlanta; Knoxville, Tennessee; Cincinnati and Cleveland, Ohio; and Buffalo and Schenectady, New York; and continue southbound along the Hudson River to Jersey City. Note that many factors influence the route choice by railroad companies for a particular shipment. This alternative case is, therefore, presented only to illustrate the flexibility of THREAT.
- *Least-population-at-risk alternative*—The route with the least cumulative exposure to population while avoiding Washington, D.C. This route coincides with the shortest-distance alternative route from Jacksonville to the West Virginia state line. Instead of turning east toward Baltimore, this route heads north into Canada and then turns east toward Buffalo. From this point, it coincides with the most-likely alternative route and travels through Mohawk Valley, turning south at Schenectady, and passes through Hudson Valley to reach Jersey City.

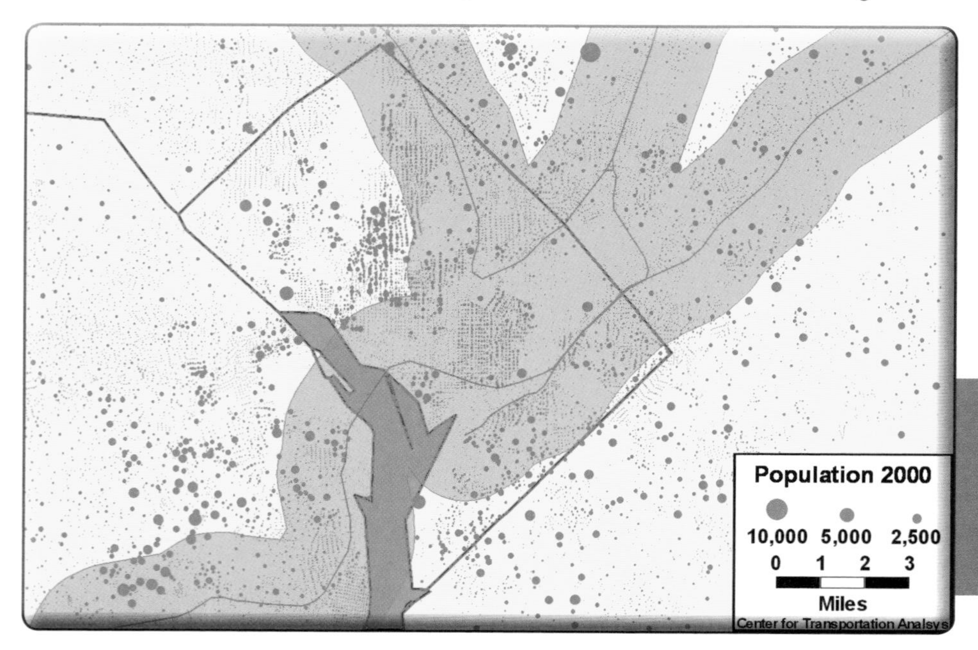

Figure 7.31 Dot-density map shows the population at risk in the exposure zones near Washington, D.C

Visualization

THREAT also has a visualization module to aid the user in comprehending competitive and interrelated routing factors, such as in-transit time, distances traversed, populations at risk, and population densities, when examining the alternatives. This visualization function also helps policy makers and officials to evaluate alternatives and communicate their decisions to the public.

The visualization module animates shipments moving on alternative routes and compares them with the base-case route—the only one that passes through the CEZ for this study. Moving dots of different colors and varying sizes are used to represent alternative routes. These dots are updated and refreshed every two minutes of in-transit time, instead of actual train-travel time. Because train scheduling depends on many factors, it is unlikely that actual train-travel time could be used for animation. Instead, the in-transit time is used in this tool as a surrogate time measure. The cumulative in-transit time is displayed in the format of days, hours, and minutes (DD:HH:MM) to track the passage of time since the departure of the shipment from its origin, excluding dwelling time in rail yards.

To afford the user a better geographical appreciation of the routes, names of major cities are displayed briefly when dots, representing shipments, traverse through cities. Cumulative at-risk population measures along these routes are fashioned like bar charts in corresponding color icons, each one representing a total of one hundred twenty-five thousand people. The size of the dots is proportional to the population densities within the specified exposure zone—three miles across for this study. The population density along the tracks of these routes varies from fifty to 17,000 people per square mile. The area with the highest population density is Jersey City.

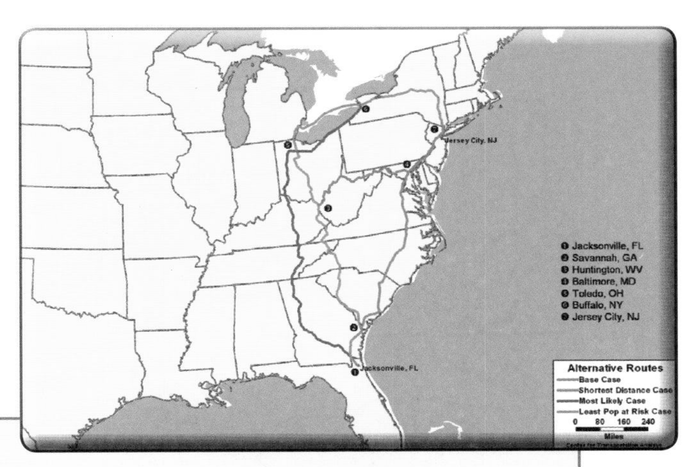

Figure 7.32 Four alternative routing scenarios are used in the case study.

This visualization function also helps policy makers and officials to evaluate alternatives and communicate their decisions to the public.

Results

Operational and performance measures generated by THREAT for the routing alternatives are tabulated. For each route, they include the overall-route distance, overall time a shipment is in transit, population along the alternative route, and the highest population density spots along the route. The percentage changes in the operational measures between each of the three alternative routes and the base case are plotted. All alternative routes, even the shortest-distance route, incur significantly extra travel distance (in the range of 56 percent–83 percent) and in-transit time (in the range of 62 percent–93 percent) in comparison with the base case.

On the other hand, the amount of population at risk saw relatively less drastic changes with a 7 percent decrease for the shortest-distance route, a 9 percent increase for the most likely alternative, and a near 20 percent drop for the least-population-at-risk route, which would have a still larger 25 percent drop if only U.S. population were considered. Obviously, all population has to be considered; and by routing hazmat shipments into Canada, certain international implications may be caused due to local (Washington, D.C.) legislative actions.

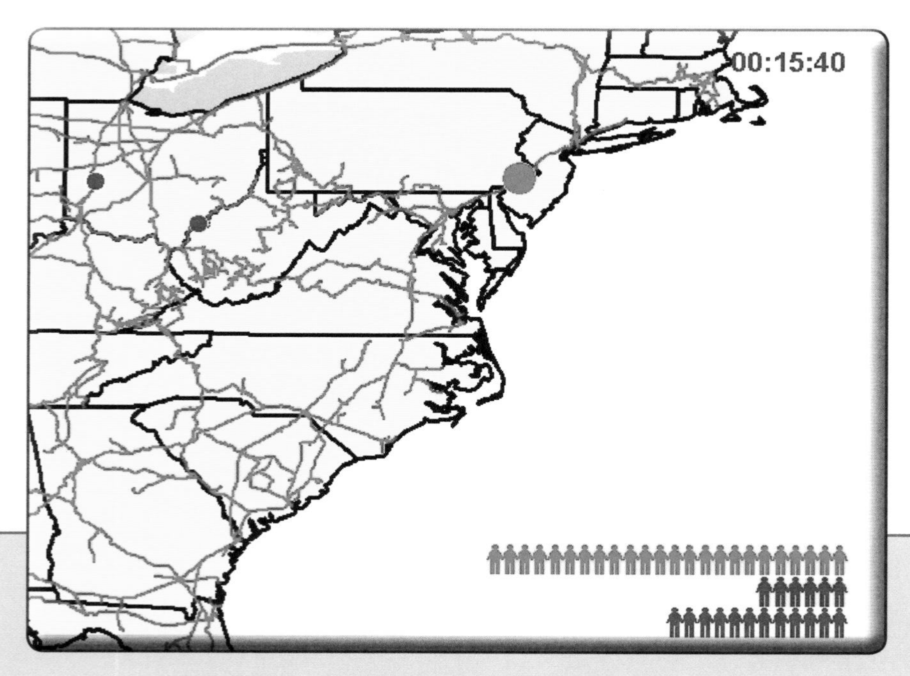

Figure 7.33 Route comparisons are animated in THREAT.

Alternative	Distance (mile)	Time in transit (hour)	Population at risk (×1,000)	Population at risk (per mile)	Highest population density spot (person/km2)
Base case	987	34	4,178	4,233	17,010 (Philadelphia)
Shortest distance	1,532	56	3,799	2,480	17,010 (Philadelphia)
Most likely	1,751	60	4,503	2,572	17,300 (north Jersey City)
Least population at risk	1,813	66	3,368 (3,147 in U.S.)	1,858	17,300 (north Jersey City)

Table 7.2 Rerouting hazmat shipments to avoid the Washington, D.C., Capitol exclusion zone incurs extra travel time and distance while not significantly reducing the at-risk population along the way.

These results seem to support certain arguments against the hazmat transportation CEZ—for the OD pair in this study and for other OD pairs currently routing through Washington, D.C. A significant increase in travel distance and time will result from CSX Transportation's efforts to avoid the CEZ. Rerouting hazmat shipments will not effectively reduce the total at-risk population along the shipment routes, but only shift the potential risk from Washington, D.C., to somewhere else.

A closer examination of the results in the form of population-density distribution profiles suggests two groups of exposure risks in high-density spots. The base case and the shortest-distance alternative did not take exposure risks into consideration. As a result, each has approximately twenty-five miles with population density higher than 10,000 people

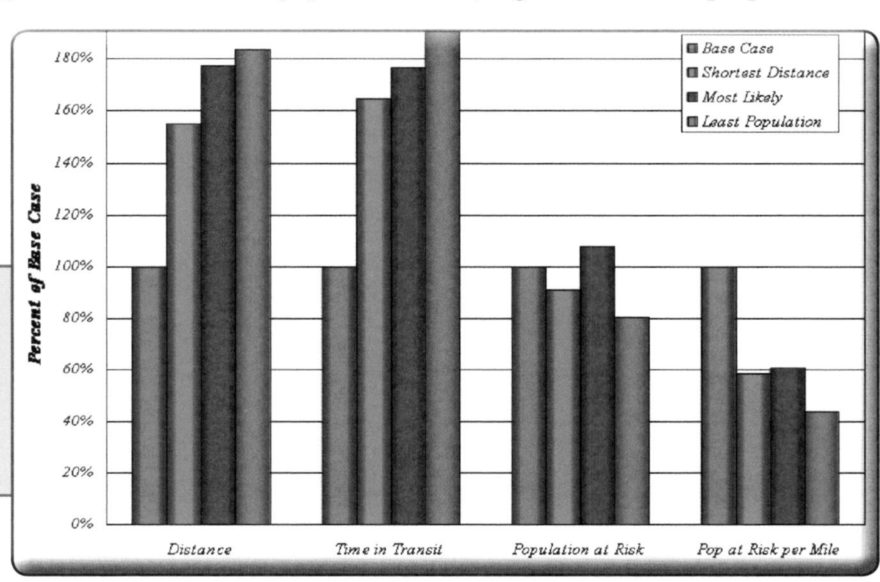

Figure 7.34 Route performance measures are shown as a percentage of the base case.

per square mile. On the other hand, the most-likely and the least-population-at-risk alternatives each has only approximately eight miles with similar population density. From the perspective of preventing terrorist attacks or sabotage, which are believed to be more likely in high-population density spots, an eight-mile area is easier to manage than a twenty-five-mile area. This suggests that the most-likely and least-population-at-risk alternatives are less attractive to terrorists and saboteurs.

The measure of effectiveness can be viewed in terms of average at-risk population per track-mile over the entire length of each route. The numbers suggest a significant decrease for all three alternative routes. This may suggest a reduced risk of exposure for day-to-day mishaps.

While Washington, D.C., is densely populated along the CSX Transportation railroad tracks, there are other cities along the CSX Transportation railroad tracks that are also densely populated, if not more so. These include Baltimore; Boston, Massachusetts; Chicago, Illinois; New York City; Jersey City; and Philadelphia. In the case that some, or all, of these cities decided to ban hazmat shipments, routing would become even more inefficient, if still possible.

Future work

The idea of using population-density distribution profiles as a measure to evaluate alternative routes suggests that an algorithm that minimizes the maximum population density along railroad tracks would be desirable. Future work in integrating a "minimax" algorithm into THREAT is being studied by the authors.

The discussion of avoiding the Washington, D.C., CEZ in this paper concentrates on finding alternative routes on railroad tracks where CSX

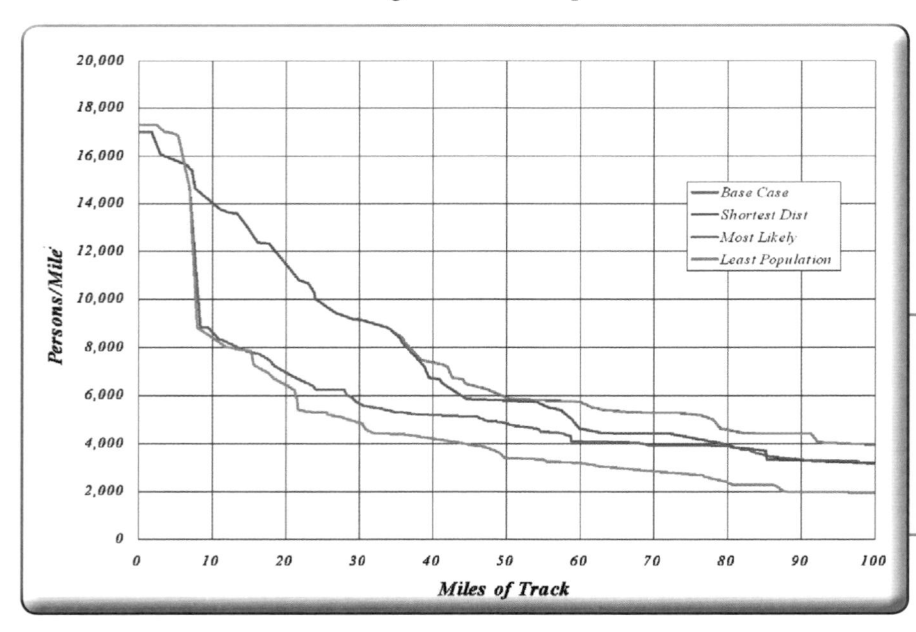

Figure 7.35 The population-density distribution profile compares the population at risk for the four routing scenarios in question.

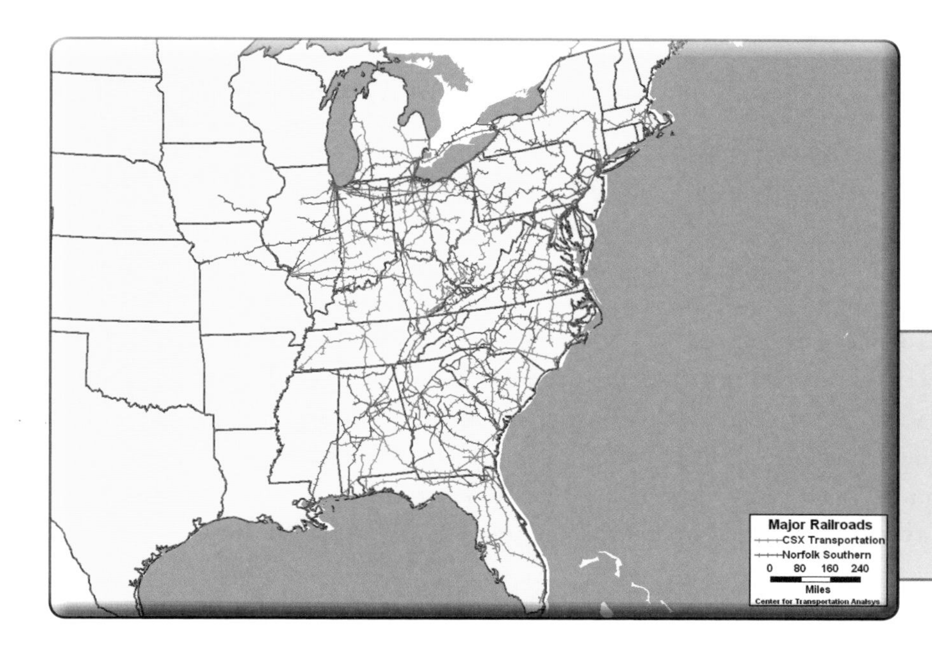

Figure 7.36 CSX Transportation and Norfolk Southern rail networks both traverse the eastern United States.

Transportation has track rights. Strategies such as inclusion of tracks from other railroads can also be evaluated using THREAT. For example, there could be more desirable alternative routes from Jacksonville to Jersey City bypassing the Washington, D.C., CEZ if certain Norfolk Southern tracks could be used. However, alternative routes involving cooperation among multiple railroad companies can be complicated.

References

Hazmat Incidents Involving Crashes or Train Derailments, Bureau of Transportation Statistics, Washington, D.C., 2005.

Leonnig, C.D., "Judge Upholds Hazmat Rail Ban—CSX Vows Appeal to Fight D.C. Law," Washington Post, Washington, D.C., April 19, 2005, B2.

Census 2000, U.S. Bureau of Census, www.census.gov/main/www/cen2000.html, accessed July 31, 2005.

Exercise in decision support

How can a community make green decisions?

The environmental movement has long touted the benefits of acting responsibly about the earth. Activities and campaigns have involved public education about all things environmentally friendly, or "green," and are combined with efforts to mitigate poor environmental stewardship. Citizens and businesses have become aware of the importance of recycling, hazardous-waste removal, cleaning up our coastlines, and reducing air pollution through sound transportation policies.

The public's awareness of one of the greatest environmental challenges facing the world today—global warming—has grown via the efforts of former Vice President Al Gore. Gore wants to alert people to the seriousness of the global-warming crisis; and with his movie, *An Inconvenient Truth,* a book of the same title, and his traveling slide show, Gore is trying to reach as many people as possible with the message.

People are beginning to ask community leaders questions about their town's sustainability policies: What is our community's carbon footprint? What is the value of a tree? What does it cost to go green?

The GIS difference

After years of global neglect, time is not on the side of officials grappling with how to implement green policies. How does a government maximize opportunities to improve the environment and lessen the community's carbon footprint? How do governments evaluate green opportunities and prioritize them based on the greatest return on investment?

In addition to being an excellent tool for visualization, GIS technology supports analysis, a use that is rooted in science. CITYgreen is an ArcGIS extension with an opportune application. It was created by American Forests, the oldest nonprofit citizens conservation organization which has led the way in developing new approaches to conservation. The first step in American Forests' plan to encourage a grass-roots approach to public policy is for communities to assess their tree-canopy cover, which is a proven indicator of a healthy and sustainable urban ecosystem. Informed decisions that take the value of natural resources into account can save cities money and help to integrate the built environment with the natural environment.

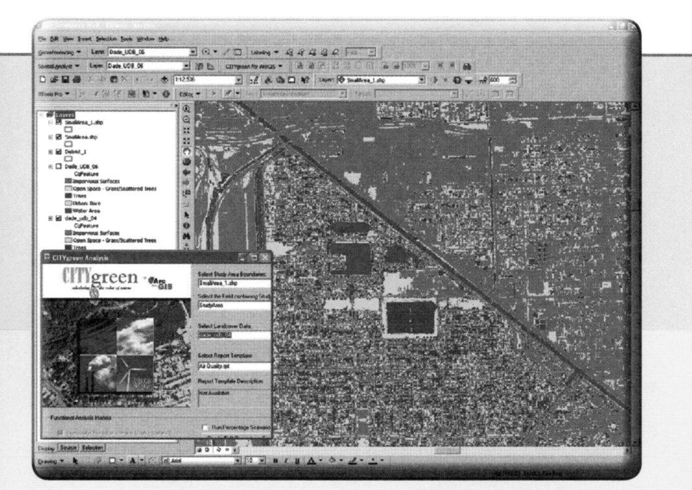

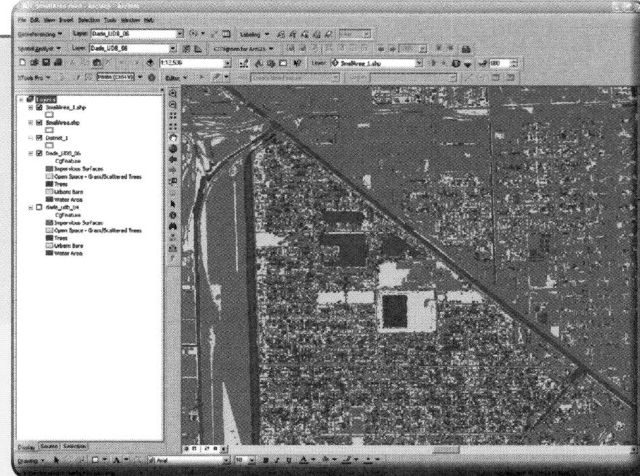

Figures 7.37 and 7.38 This neighborhood in Miami has undergone a significant land-cover change between 2004 (left) and 2006 (right) as a result of hurricanes and land development, with a 14 percent loss in its tree canopy. Even though this is a small area and the changes are relatively minor, the effects on the environment have been significant.

To help cities uncover the hidden values that trees and green space provide, American Forests developed a GIS analysis technique called the Regional Ecosystem Analysis. The process uses satellite data, field surveys, and GIS software including CITYgreen.

CITYgreen software conducts complex analyses of ecosystem services, creates easy-to-understand reports, and calculates the dollar benefits of tree-related services in specific areas. CITYgreen supports tree-ordinance modeling, carbon-offset calculations, storm-water-runoff modeling, air-pollution removal, and land-use modeling.

For more information on the American Forests CITYgreen application, visit www.americanforests.org/productsandpubs/citygreen/.

AMERICAN FORESTS®
americanforests.org

Analysis Report

Conditions	Conditions
2004	2006

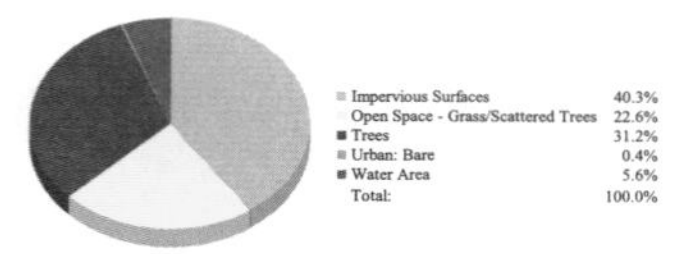

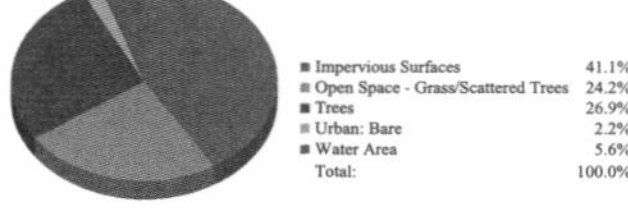

Air Quality Results

Pounds Removed per Year

Pollutant	2004	2006
Carbon Monoxide:	640	552
Nitrogen Dioxide:	2,175	1,878
Ozone:	7,036	6,075
Particulate Matter:	5,885	5,081
Sulfer Dioxide:	512	442
Total:	**16,247**	**14,028**

By absorbing and filtering out nitrogen dioxide (NO2), sulfur dioxide (SO2), ozone (O3), carbon monoxide (CO), and particulate matter less than 10 microns (PM10) in their leaves, urban trees perform a vital air cleaning service that directly affects the well-being of urban dwellers. This model, UFORE, developed the the US Forest Service, estimates the annual air pollution removal rate of trees within a defined study area for the pollutants listed below. To calculate the dollar value of these pollutants, economists use "externality" costs, or indirect costs borne by society such as rising health care expenditures and reduced tourism revenue. The actual externality costs used in the model is set by the each state, Public Services Commission.

Benefits Summary

Landcover Change (acres)			
Landcover	**2004**	**2006**	**Change**
Tree Canopy:	144	124	-14%
Air Pollution Benefits			
Pollutants Removed (lbs):	16,247	14,028	-2,220
$ Amount:	$41,027	$35,422	-$5,605
Carbon Stored (tons):	6,176	5,332	-844
Carbon Sequestered (lbs):	48	42	-7

Stormwater Results

Stormwater Volume Change Summary

2-yr, 24-hr Rainfall: 6.00 in.	
**Curve Number reflecting 2004 conditions:*	82
**Curve Number reflecting 2006 conditions :*	83
Change in stormwater volume due to landcover change:	174,302 cu. ft.
Construction cost, per cu. ft.of stormwater, to build retention facility :	$2.00
Cost of stormwater retention resulting from landcover change:	**$348,604**

Water Quality (Contaminant Loading)

Percent Change in Contaminant Loadings from 2004 conditions to 2006 conditions

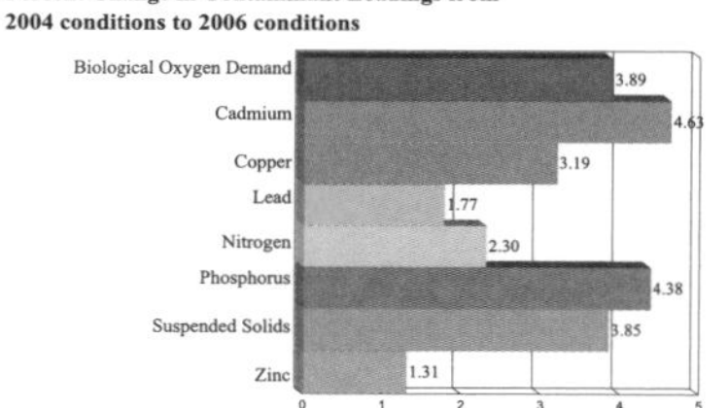

Notes: *The stormwater calculations are based on curve number which is an index developed by the NRCS, to represent the potential for storm water runoff within a drainage area. Curve numbers range from 30 to 100. The higher the curve number the more runoff will occur. The change in curve number reflects the increase/decrease in the volume of stormwater runoff.

Figure 7.39 CITYgreen software was created by American Forests to encourage conservation through its analysis of changes in the tree canopy.

Image credits

Figure 1.1 Courtesy of City of Newark, New Jersey.
Figure 1.2 Courtesy of City of Newark, New Jersey.
Figure 1.3 Courtesy of City of Newark, New Jersey.
Figure 1.4 Courtesy of City of Newark, New Jersey.
Figure 1.5 Courtesy of City and County of Denver, Colorado.
Figure 1.6 Courtesy of City and County of Denver, Colorado.
Figure 1.7 Courtesy of RouteSmart Technologies Inc.
Figure 1.8 Courtesy of RouteSmart Technologies Inc.
Figure 1.9 Courtesy of Thinkstock.
Figure 2.1 Courtesy of City of Frederick, Maryland.
Table 2.1 Courtesy of City of Frederick, Maryland.
Figure 2.2 Courtesy of City of Frederick, Maryland.
Table 2.2 Courtesy of U.S. National Park Service.
Figure 2.3 Courtesy of City of Frederick, Maryland.
Figure 2.4 Courtesy of City of Houston, Texas.
Figure 2.5 Courtesy of City of Houston, Texas.
Figure 2.6 Courtesy of City of Houston, Texas.
Figure 2.7 Courtesy of City of Houston, Texas.
Figure 2.8 Courtesy of U.S. National Park Service.
Figure 2.9 Courtesy of U.S. National Park Service.
Figure 2.10 Courtesy of U.S. National Park Service.
Figure 2.11 Courtesy of U.S. National Park Service.
Figure 2.12 Courtesy of U.S. National Park Service.
Figure 2.13 Courtesy of U.S. National Park Service.
Figure 2.14 Courtesy of U.S. National Park Service.
Figure 2.15 Courtesy of U.S. National Park Service.
Figure 2.16 Courtesy of U.S. National Park Service.
Figure 2.17 Courtesy of U.S. National Park Service.
Figure 2.18 Courtesy of Oneida Nation, Oneida, Wisconsin.
Figure 2.19 Screenshot by ESRI, StreetMap USA data courtesy ESRI.
Figure 2.20 Screenshot by ESRI, StreetMap USA data courtesy ESRI.
Figure 2.21 Courtesy of ESRI.
Figure 2.22 Courtesy of City of Moreno Valley, California.
Figure 2.23 Courtesy of City of Moreno Valley, California.
Figure 3.1 Courtesy of City of Cleveland, Ohio.
Figure 3.2 Courtesy of City of Cleveland, Ohio.
Figure 3.3 Courtesy of City of Cleveland, Ohio.
Figure 3.4 Courtesy of City of Cleveland, Ohio.
Figure 3.5 Courtesy of City of Cleveland, Ohio.

Figure 3.6 By Buck Bennett, courtesy of Department of Natural Resources, Coastal Resources Division, Brunswick, Georgia.
Figure 3.7 By Jim Couch, Georgia Parks, courtesy of Georgia Department of Natural Resources, Brunswick, Georgia.
Figure 3.8 Courtesy of Georgia Department of Natural Resources, Brunswick, Georgia.
Figure 3.9 Courtesy of Shawn Jordan, Department of Natural Resources, Coastal Resources Division, Brunswick, Georgia.
Figure 3.10 Courtesy of Georgia Department of Natural Resources, Brunswick, Georgia.
Figure 3.11 Courtesy of Georgia Department of Natural Resources, Brunswick, Georgia.
Figure 3.12 By Buck Bennett, courtesy of Department of Natural Resources, Coastal Resources Division, Brunswick, Georgia.
Figure 3.13 Courtesy of Philadelphia GIS Services Group, Philadelphia, Pennsylvania.
Figure 3.14 Courtesy of Philadelphia GIS Services Group, Philadelphia, Pennsylvania.
Figure 3.15 Courtesy of Philadelphia GIS Services Group, Philadelphia, Pennsylvania.
Figure 3.16 Courtesy of Philadelphia GIS Services Group, Philadelphia, Pennsylvania.
Figure 3.17 Courtesy of Philadelphia GIS Services Group, Philadelphia, Pennsylvania.
Figure 3.18 Courtesy of Washington County Planning and Parks Department, West Bend, Wisconsin.
Figure 3.19 Courtesy of Washington County Planning and Parks Department, West Bend, Wisconsin.
Figure 3.20 Courtesy of City of Cleveland, Ohio.
Figure 3.21 Courtesy of Jim Ollerton.
Figure 3.22 Screenshot by ESRI, data from Riverside County, AirPhoto USA.
Figure 3.23 Screenshot by ESRI, data from Riverside County, U.S. Census Bureau.
Figure 3.24 Photo by Ryan McVay, courtesy of Photodisc/Getty Images.
Figure 3.25 Screenshot by ESRI.
Figure 3.26 Screenshot by ESRI, featuring ArcGIS Online Beta.
Figure 4.1 Courtesy of Fort Bend County, Texas.
Figure 4.2 Courtesy of Fort Bend County, Texas.
Figure 4.3 Courtesy of Miami-Dade Office of Emergency Management, Miami, Florida.
Figure 4.4 Courtesy of Miami-Dade Office of Emergency Management, Miami, Florida.
Figure 4.5 Courtesy of Miami-Dade Office of Emergency Management, Miami, Florida.
Figure 4.6 Courtesy of Peterson Air Force Base, Colorado Springs, Colorado.
Figure 4.7 Photo by Purestock, courtesy of Purestock/Getty Images.
Figure 4.8 Photo by C. Lee/PhotoLink, courtesy of Photodisc/Getty Images.
Figure 4.9 Courtesy of ESRI.
Figure 4.10 Photo by Bruce Weniger, courtesy of CDC Public Health Image Library.
Figure 4.11 Photo by PhotoLink/Photodisc/Getty Images.
Figure 5.1 Courtesy of City and County of Denver, Colorado.
Figure 5.2 Courtesy of City and County of Denver, Colorado.
Figure 5.3 Courtesy of City and County of Denver, Colorado.
Figure 5.4 Courtesy of City and County of Denver, Colorado.
Figure 5.5 Courtesy of City of Pasadena, California.
Figure 5.6 Courtesy of City of Pasadena, California.
Figure 5.7 Courtesy of City of Pasadena, California.
Figure 5.8 Courtesy of U.S. Geological Survey.
Figure 5.9 Courtesy of U.S. Geological Survey.
Figure 5.10 Courtesy of U.S. Geological Survey.
Figure 5.11 Courtesy of U.S. Geological Survey.
Figure 5.12 Courtesy of Utah Department of Health.
Figure 5.13 Courtesy of Utah Department of Health.
Figure 5.14 Courtesy of Utah Department of Health.
Figure 5.15 Courtesy of ESRI.
Figure 5.16 Photo by Comstock, copyright 2008 Jupiter Images Corporation.
Figure 6.1 Courtesy of City of Jacksonville, Florida.
Figure 6.2 Courtesy of City of Jacksonville, Florida.
Figure 6.3 Courtesy of City of Jacksonville, Florida.
Figure 6.4 Courtesy of City of Jacksonville, Florida.

Figure 6.5 Courtesy of Pottawattamie County, Iowa.
Figure 6.6 Courtesy of Pottawattamie County, Iowa.
Figure 6.7 Courtesy of Pottawattamie County, Iowa.
Figure 6.8 Courtesy of Zanesville/Muskingum County Health Department, Zanesville, Ohio.
Figure 6.9 Courtesy of Zanesville/Muskingum County Health Department, Zanesville, Ohio.
Figure 6.10 Courtesy of Zanesville/Muskingum County Health Department, Zanesville, Ohio.
Figure 7.1 Courtesy of Curry County GIS, Curry County, Oregon.
Figure 7.2 Courtesy of Curry County GIS, Curry County, Oregon.
Figure 7.3 Courtesy of Georgia Department of Human Resources, Division of Public Health, Atlanta, Georgia.
Table 7.1 Courtesy of Georgia Department of Human Resources, Division of Public Health, Atlanta, Georgia.
Figure 7.4 Courtesy of Georgia Department of Human Resources, Division of Public Health, Atlanta, Georgia.
Figure 7.5 Courtesy of Georgia Department of Human Resources, Division of Public Health, Atlanta, Georgia.
Figure 7.6 Courtesy of Georgia Department of Human Resources, Division of Public Health, Atlanta, Georgia.
Figure 7.7 Courtesy of Georgia Department of Human Resources, Division of Public Health, Atlanta, Georgia.
Figure 7.8 Courtesy of Idaho Department of Environmental Quality.
Figure 7.9 Courtesy of Idaho Department of Environmental Quality.
Figure 7.10 Courtesy of Idaho Department of Environmental Quality.
Figure 7.11 Courtesy of Idaho Department of Environmental Quality.
Figure 7.12 Courtesy of Idaho Department of Environmental Quality.
Figure 7.13 Courtesy of Metropolitan Council, St. Paul, Minnesota.
Figure 7.14 Courtesy of Metropolitan Council, St. Paul, Minnesota.
Figure 7.15 Courtesy of Metropolitan Council, St. Paul, Minnesota.
Figure 7.16 Courtesy of Preservation League of New York State.
Figure 7.17 Courtesy of Preservation League of New York State.
Figure 7.18 Courtesy of Preservation League of New York State.
Figure 7.19 Courtesy of Preservation League of New York State.
Figure 7.20 Courtesy of Preservation League of New York State.
Figure 7.21 Courtesy of Preservation League of New York State.
Figure 7.22 Courtesy of City of Sheridan Planning Division, Sheridan, Wyoming.
Figure 7.23 Courtesy of City of Sheridan Planning Division, Sheridan, Wyoming.
Figure 7.24 Courtesy of City of Sheridan Planning Division, Sheridan, Wyoming.
Figure 7.25 Courtesy of City of Sheridan Planning Division, Sheridan, Wyoming.
Figure 7.26 Courtesy of City of Sheridan Planning Division, Sheridan, Wyoming.
Figure 7.27 Courtesy of Oak Ridge National Laboratory.
Figure 7.28 Courtesy of Oak Ridge National Laboratory.
Figure 7.29 Courtesy of Oak Ridge National Laboratory.
Figure 7.30 Courtesy of Oak Ridge National Laboratory.
Figure 7.31 Courtesy of Oak Ridge National Laboratory.
Figure 7.32 Courtesy of Oak Ridge National Laboratory.
Figure 7.33 Courtesy of Oak Ridge National Laboratory.
Table 7.2 Courtesy of Oak Ridge National Laboratory.
Figure 7.34 Courtesy of Oak Ridge National Laboratory.
Figure 7.35 Courtesy of Oak Ridge National Laboratory.
Figure 7.36 Courtesy of Oak Ridge National Laboratory.
Figure 7.37 Courtesy of American Forest.
Figure 7.38 Courtesy of American Forest.
Figure 7.39 Courtesy of American Forest.

Index